BBC

Springwatch

BRITISH
WILD
LIFE

*To Britain's wonderful wildlife, for giving
so much pleasure to so many people.*

BBC Springwatch

BRITISH WILD LIFE

Stephen Moss
and the
Springwatch Team

Collins

HarperCollins Publishers
77-85 Fulham Palace Road
London W6 8JB

www.harpercollins.co.uk

Collins is a registered trademark
of HarperCollins Publishers Ltd.

First published in 2012

By arrangement with the BBC
The BBC logo is a trademark of the British
Broadcasting Corporation and is under licence.

BBC logo © BBC 1996

17 16 15 14 13 12
10 9 8 7 6 5 4 3 2 1

A catalogue record for this book is available from the British Library.

ISBN 978-0-00-746286-5

Colour reproduction by FMG
Printed and bound in Italy by L.E.G.O. SpA

Contents

Introduction

Like the monarchy, afternoon tea and the weather, *Springwatch* is a British institution. Of course it didn't start that way: in fact, when it began, back in 2003, it wasn't called *Springwatch* at all. *Wild in Your Garden* was a week of live TV programmes broadcast from a suburb of Bristol, presented by Bill Oddie, Kate Humble and Simon King. The following year it moved to Devon, was extended to three weeks, and turned into *Britain Goes Wild*. It wasn't until 2005 that it finally acquired the name *Springwatch*.

Since then, it has changed location twice (to Pensthorpe in Norfolk and Ynys-Hir in mid-Wales), spawned numerous spin-offs, including *Autumnwatch*, *Snow Watch* and *Winter Watch*, and acquired a whole range of new presenters including Chris Packham, Martin Hughes-Games and Michaela Strachan. It has also become as fixed a part of the TV schedules as *EastEnders* and *Top Gear*.

The success of all these shows, and the various website content, campaigns and events surrounding them, is reflected in the fact that the British have rediscovered their age-old passion for wildlife. Actually it was always there, it was just that for a long time those in charge of the TV schedules preferred their wildlife to be foreign and exotic – lions and elephants rather than badgers and blue tits. *Springwatch* has shown that there is a real appetite not just for watching wildlife on our doorstep, but for learning about it, too.

That is what the *Springwatch Guide to British Wildlife* is all about. It is designed to help you learn more about the wildlife featured in the programmes: from the common and familiar species such as blackbirds and blue tits, through the more obscure examples such as natterjack toads and rare orchids, to iconic creatures such as the killer whale and white-tailed eagle. Each creature or group of plants or animals gets its own illustrated double-page spread, featuring fascinating facts and details of their lives.

Of course this can only scratch the surface: there are plenty of specialised guides available to every group of Britain's wild creatures. But once you are hooked, it will hopefully inspire you to learn more about the wildlife we treasure so much.

The white-tailed eagle (also known as the sea eagle) is Britain's largest bird of prey, with a wingspan of over 2m (6ft).

Birds

Tits

Of all the stars of *Springwatch*, perhaps none is quite so enduringly popular as the blue and great tits. Their antics, as they jostle for position on feeders, their cheeky behaviour and, above all, their life-or-death struggles as they bring up a brood of chicks under the glare of TV cameras, make them utterly compelling to millions of viewers.

There is one simple reason why these birds have become such stars: they choose to nest in bird boxes, which means that from the earliest days of live TV broadcasts – certainly as far back as the early 1990s and *Bird in the Nest* – it was possible to put cameras into their nests. Things have changed a lot since then, of course: instead of huge cameras, we now have a battery of tiny high-definition cameras that can view the nesting birds at all angles, without causing them a moment's disturbance.

Blue tits (above) and great tits (below) are originally woodland birds, which have readily adapted to life in our gardens.

The reason tits take so readily to nest boxes is that these artificial homes are an ideal substitute for the natural holes and crevices in which these birds nest. For, like so many of our garden birds, tits are originally birds of the woodland or woodland edge. They have made their homes in our gardens primarily because we provide plenty of food for them – not just in the shape of seeds and peanuts but also through what we plant, which provides much-needed moth caterpillars for their hungry broods.

And broods don't get much hungrier than a dozen baby blue tits, all crammed into their soft, cup-shaped nest of moss and feathers, inside a tiny wooden box. During the dozen or so days between the chicks hatching and when they finally fledge and leave the nest, a single brood may require as many as 12,000 caterpillars – that's 1,000 caterpillars per day, or roughly one caterpillar every two minutes for each of the two parents. No wonder the adult birds look so tatty and exhausted by the end of the breeding season!

Blue and great tits are the two species we often feature on *Springwatch*, along with occasional appearances by a third garden species, the coal tit. These three are easy to tell apart: the great tit, as its name suggests, is larger than the other two, and has a smart black and yellow plumage with a green back, white cheeks and a thick black strip down its front. Blue tits are smaller, mainly blue and yellow, and also with white cheeks (though the youngsters have yellow cheeks). Coal tits are basically a monochrome version of the blue tit

– similar in size, with a black, brown and grey plumage and a distinctive white patch on the back of the neck, or nape.

Three other kinds of tit breed in Britain, two of which, the marsh and willow, are so similar that they weren't separated as different species in Britain until the last year of the nineteenth century. Both are superficially similar to the coal tit, being mainly brown with a black cap, but they are both larger and less varied in plumage, and lack the coal tit's white nape. Sadly, both marsh and willow tits have suffered major declines in recent years, perhaps due to a loss of suitable breeding habitat.

The final species, the crested tit, is found only in the ancient pine forests of Speyside in Scotland, and can easily be identified by its perky crest. Two other species with 'tit' in their name, the long-tailed tit and the bearded tit, are in fact unrelated to the tits, though long-tailed tits do often accompany flocks of tits in winter.

Unlike many other nesting birds featured on *Springwatch*, both blue and great tits usually have only one brood of chicks in a season. As they generally live for only a year or two as adults, this is a risky strategy: literally putting all their eggs in one basket. In years when spring comes very early, as in 2007 and 2011, this can mean that by the time the birds begin to nest, the moth caterpillars on which they so depend are already coming to the end of their lifecycle and beginning to pupate, which can mean very bad years for the birds. It remains to be seen whether these species are bringing their own breeding season significantly earlier to cope with this change. Evidence suggests that they probably are.

Once their chicks are safely out of the nest and have fledged, adult tits appear to vanish for a month or so; they are moulting into a bright new plumage and hide away to avoid predators. By autumn, they appear again, teaming up with those of the same and other species in mixed flocks, which rove woodlands or gardens, uttering the tiny, high-pitched contact calls that draw our attention to them. They travel in circuits in search of food; it has been calculated that if you see half a dozen blue tits coming to your feeder during the course of a winter's day, you may in fact play host to as many as ten times that number. So make sure you keep your feeders clean and well stocked with food!

Tits such as coal (top left), crested (above) and willow (below) use their sharp, pointed bills to feed on a range of insects and seeds.

Finches

Of all our birds, there could hardly be a more colourful, attractive and better-known family than our finches. The dozen or so species of finch that regularly breed in Britain include some of our most familiar garden birds – the chaffinch, goldfinch and greenfinch – as well as some scarce and elusive ones such as the hawfinch and crossbill. Other species include the siskin and redpoll, which are growing in familiarity and now visit gardens more than they did in the past, and sadly declining species such as the bullfinch, linnet and twite.

Finches are seed-eaters, which means they have taken well to hanging feeders and bird tables containing either seeds or peanuts. Each species has subtly different food requirements, often leading to different-shaped bills. So, whereas the goldfinch has a slender, pointed bill, which it uses to extract the tiny seeds from thistles and teasels (and to take nyjer seed from garden feeders), the bullfinch has a much thicker, sturdier bill, which it uses to feed on buds. The hawfinch – the biggest British finch – sports a massive bill, which it uses to crack cherry stones, applying a pressure of more than two tons per square inch!

But of all the finches, by far the most bizarre feeding adaptation comes in the crossbills, whose various species have uniquely developed crossed mandibles: one half of their bill crosses over the other, enabling the birds to extract the seeds from pine cones and the fruits of other coniferous trees. Because they don't depend on insects or fruit to feed their young, they nest remarkably early – sometimes beginning at the start of the new year. The crossbills are our only nomadic breeding bird, with flocks shifting around the country (and occasionally disappearing across the North Sea to Scandinavia) from year to year. The group also includes Britain's only endemic breeding bird, the Scottish crossbill, which is confined to a tiny area of the Caledonian pine forests of northeast Scotland.

Our commonest finch by far is the chaffinch, which, with more than five-and-a-half million breeding pairs, is second only to the wren as Britain's commonest breeding bird. Chaffinches are found throughout the UK but are especially common in Scottish

The chaffinch is our commonest finch, found in a wide range of wooded and garden habitats.

The goldfinch is currently on the increase, thanks partly to our habit of feeding birds in our gardens.

woodlands, where vast flocks gather to feed in winter. But, like other woodland birds, they have adapted well to living alongside us in our gardens.

Two other species – the greenfinch and the goldfinch – are even more dependent on garden bird-feeding; indeed, the goldfinch population has risen dramatically in the past couple of decades thanks to us providing high-energy food such as sunflower hearts. Today, the tinkling sound of a goldfinch flock can be heard in many gardens. Siskins – a streaky, smaller version of the greenfinch – have also learned to take advantage of our generosity, and have spread southwards from Scotland to southern England as a result.

One problem, though, has come with this success. Being sociable birds, often coming to gardens in flocks, finches are especially susceptible to disease, with greenfinches being hit particularly hard. It can be distressing to find dead birds underneath your feeders, so make sure you clean them thoroughly, and remove any mouldy food.

Some kinds of finch hardly ever come to gardens, and they are perhaps suffering as a result. Linnet and twite – two small, streaky finches with subtly beautiful plumage features – have both declined dramatically, and are now on the Red List of birds of conservation concern. Linnets are farmland birds, and changes in agricultural practices, such as the sowing of winter wheat, which has reduced the amount of stubble available for the birds to feed on in winter, have contributed to their decline. The decline of the twite is, in some ways, even more worrying; this moorland bird has virtually disappeared from northern England and now lives mainly in Scotland.

Finches build their open, cup-shaped nests hidden away in bushes, trees and shrubs. They mostly lay between four and six eggs, and usually raise two or even three broods of chicks during a single breeding season.

All finches, including greenfinch (top left), bullfinch (bottom left), crossbill (above) and siskin (below), have developed specially shaped bills to suit their favoured foods.

Buntings

Buntings are, in some ways, the Cinderellas of seed-eating birds. They are neither as colourful as the finches, nor as ubiquitous as the sparrows, and we rarely see them unless we are looking specifically for them. This is partly because, until recently, they were mainly birds of the open countryside, but they have also suffered from changes in the way the countryside is farmed.

Five species of bunting breed in Britain: the corn bunting, cirl bunting and yellowhammer, all found mainly on lowland farms; the reed bunting, found in wetland habitats in the breeding season, and farms and gardens in autumn and winter; and our rarest, the snow bunting, found on the high tops of the Cairngorm mountains in summer and in flocks along our coasts in winter.

The corn bunting (above), yellowhammer (right) and cirl bunting (below) have all suffered in recent years from modern farming methods.

All five species are, sadly, giving cause for concern: the three farmland species are on the Red List, while the other two are on the Amber List. The corn bunting and yellowhammer really have seen their populations nose-dive. Once found throughout lowland Britain, they have disappeared from many of their former haunts. This is due mainly to the loss of breeding habitat and winter stubble fields caused by modern farming methods, which need to extract the maximum possible yield from fields all year round. The cirl bunting, a shy relative of the yellowhammer, once almost became extinct as a British breeding bird, but, thanks to the RSPB and farmers in its south Devon stronghold, it is now making a comeback.

Telling male buntings apart is reasonably easy, but females and youngsters can be puzzling – they are classic 'little brown jobs'! The corn bunting is one of our largest songbirds: bulky and fat, with a streaky plumage, a large head and bill, and a characteristic way of dangling its legs when it flies. It may be a fairly dull-looking bird, but the corn bunting has some extraordinary breeding habits. The males will defend a large territory, supporting several females, a strategy known as 'polygyny'.

The yellowhammer (its name comes from the German word for bunting) is easy to identify in breeding plumage: the male's custard-yellow head is very striking; and the same is true for the male reed bunting, with its very obvious black and white head pattern. Females are trickier: yellowhammers can look quite brown, while a female reed bunting is superficially like a sparrow, but streakier, especially on her head. Both male and female reed buntings also show white outer tail feathers as they fly away from you! Cirl buntings look a bit like a bird designed by a committee: the body is streaky brown like a dunnock, while the head is black and yellow, similar to that of a yellowhammer but with a greenish tinge.

Song is a good way to tell the three commoner British buntings apart: the corn bunting sounds exactly like someone rattling a bunch of keys; the yellowhammer sings the famous mnemonic 'a-little-bit-of-bread-and-no-cheeeeese'; while the reed bunting makes a rather pathetic sound, sometimes rendered as 'pick-my-liquorice', though it also sounds something like a bored sound engineer, continually repeating the mantra of 'testing... one... two... testing'.

Both yellowhammers and reed buntings now regularly come to bird tables, especially in rural areas or where there is arable farmland or a wetland nearby. This has helped them maintain their numbers at a time when the wider countryside can no longer provide the food they need, and also brought a touch of the exotic among the usual tits, finches and sparrows.

The odd man out of our buntings is the snow bunting. It doesn't really belong in Britain, as it is adapted to living in the far north, and is the only small bird to spend the winter months in the Arctic. Nevertheless, a few pairs nest each summer in snow-covered parts of the Scottish Highlands, while in winter little flocks can sometimes be seen along our east coasts. In flight, they are very easy to identify – like a flock of snowflakes scattering through the air!

The male reed bunting (left) is one of our most handsome breeding birds.

The snow bunting (below) is one of the toughest birds on the planet.

Blackbird, Song Thrush & Mistle Thrush

These are three of our most familiar birds, with distinctive and well-loved songs: the blackbird and song thrush regularly top the hit parade of Britain's favourite songsters. All three are resident, though numbers are supplemented each autumn and winter with new arrivals from across the North Sea, taking advantage of our milder climate. All three are also well adapted to gardens, though the mistle thrush is also commonly found in more open countryside and parkland, with tall trees from which it can deliver its song.

Indeed, the mistle thrush is often one of the first songbirds to sing, at the start of the new year – even earlier during mild winters. Thomas Hardy wrote a famous poem, *The Darkling Thrush*, which celebrates a thrush singing during blustery weather on New Year's Eve; many people think that this early date and the bad weather point to this being a mistle, rather than a song, thrush. Mistle thrushes have earned the folk name 'stormcock' for their habit of singing through the worst of the British weather.

The song thrush is usually the next of the trio to sing, usually starting off in January or early February. Its song is probably the most distinctive of all our birds: measured phrases, repeated in twos or threes, as if the bird were talking to you! Soon afterwards the blackbird joins in, deeper than the song thrush, with a really fluty tone.

All three birds often choose a high perch from which to deliver their songs, which makes them easy to see. This is a good time of year to get to grips with the differences between them. Male blackbirds are easy: coal-black in colour, with that distinctive bright custard-yellow bill and – if you get close enough to see it – a yellow ring around the eye. Female blackbirds, which tend to be more skulking in habit, are chocolate-brown above and below.

The two thrushes are harder to tell apart, though with practice not too tricky. The mistle thrush is much larger, paler and greyer, with heavy spotting below and a permanently surprised expression on its face. Song thrushes are smaller, plumper and a deeper brown colour, and with a much more kindly expression! Beware young blackbirds, especially in late summer, which can be a bit streaky below and a little thrush-like.

Blackbirds often feed on berry bushes in winter to supplement their energy needs.

Mistle thrush (above) and song thrush (opposite) enjoy a varied diet, including fruit as well as insects, snails and slugs.

These birds' nests are easy to find too, often tucked away in a shrub or bush or, in the case of the mistle thrush, in the fork of a tree. All three start to breed very early on, and sometimes have eggs (and occasionally even chicks) before Christmas or early in the new year. The two thrushes have two broods – the song thrush sometimes three – but blackbirds really are the champion nesters, often raising three, four or even five broods of chicks in a single breeding season. This avoids the 'putting all your eggs in one basket' problem faced by blue tits, and means that even if there is prolonged bad weather during some parts of the breeding season, they can usually get at least one brood of chicks to fledging.

The blackbird and mistle thrush eat a wide range of foods, including snails and worms as well as berries and fruit – a mistle thrush will often spend the winter defending a single berry bush against all-comers. Song thrushes are more carnivorous, feeding mainly on slugs and snails, which they bash on a heavy stone, known as an 'anvil', to get them out of their shells. Song thrushes suffer badly from poisoning by agricultural and garden pesticides, especially slug pellets – being a natural pest controller, they should be welcomed into the garden! In autumn and winter, all three species, but especially blackbirds, will feed on windfall apples.

The availability of food and places where the birds can nest means that gardens are incredibly important for both blackbirds and song thrushes – they nest there in densities up to ten times greater than in their natural woodland habitats.

Winter Thrushes & Waxwing

From time to time, as frost and snow take their grip on the British winter, three beautiful and exotic-looking visitors to our shores turn up in our gardens, to the delight of anyone lucky enough to get close views of them.

Two of these winter visitors are members of the thrush family, closely related to the more familiar song and mistle thrushes. The redwing and the fieldfare are often known simply as 'winter thrushes', as they come in their millions each autumn from Scandinavia and Iceland to spend the winter months here. Like waders and wildfowl, they do so to take advantage of our relatively mild winter climate compared with places to the north and east of Britain.

Usually they spend most of the time roaming the wider countryside in huge flocks, stripping the hedgerows bare of berries or feeding in muddy fields. But when snow and ice make finding food more difficult, they often head into our gardens in search of windfall apples, berry bushes and other food we provide.

Of the two, the redwing is by far the most familiar. Our smallest thrush, it is a shade shorter than the song thrush, with darker brown upper parts, spotted under parts, and a distinctive creamy stripe above the eye. Its name comes from the orange-red patch on its flanks, which is more obvious as the bird takes to the air, lifting its wings.

Fieldfare (above) and redwing (below) are two of the commonest species wintering in Britain.

The fieldfare is even more colourful: a large bird, almost as big as a mistle thrush, with a long tail, grey head, reddish-brown back and warm yellow on the breast. They look very different from the usual garden birds, so much so that when they invaded our gardens a couple of winters ago, there were several reports of 'cuckoos' in the middle of winter! Like mistle thrushes, fieldfares can be aggressive birds, defending a berry bush against any intruders that might steal their food. But they are less solitary than our two resident species, and usually travel in noisy flocks, chattering to each other as they go.

Both redwings and fieldfares arrive in Britain during October, the redwings usually a week or two ahead of their larger cousins. On clear autumn nights, listen out for their distinctive high-pitched call, which people once supposed was made by witches flying overhead on their broomsticks!

Redwings and fieldfares can be seen in the countryside right the way through to March or even April, but once spring has arrived, the vast majority have already left our shores: the fieldfares to Scandinavia and northern Russia; the redwings to Scandinavia and Iceland. Very few remain to breed, which in some ways is odd, given the huge numbers wintering here, and the fact that fieldfares, in particular, can be found nesting as far south as Hungary, on the same latitude as southern Britain.

The third member of this winter trio is even more exotic than the other two. Waxwings are starling-sized birds with a delicate buffish-brown plumage, black and yellow wings tipped with red – these resemble sealing wax, hence the bird's name – and a wispy crest, giving them a faintly comical appearance.

Waxwings may be rare and exotic, but they are often found in gardens where they feed on berry bushes.

Unlike other winter visitors, waxwings are a so-called 'irruptive' species, which means that in some years hardly any come to spend the winter in Britain, while in other years there may be tens of thousands. Their appearance here does not, as some suppose, foretell a hard winter. It is simply a reflection of their population level and the amount of food back home in Scandinavia – during years of berry shortages and high numbers of birds, they will head south, otherwise they stay put.

Unlike other rare birds, waxwings rather like our suburbs and gardens, where they can often find plenty of juicy red berries to eat. They also regularly turn up in supermarket car parks, which are often planted with berry-bearing bushes whose fruit appears at just the right time for the waxwings' arrival.

If you are lucky enough to discover a flock of waxwings feeding in your front garden, be prepared for another invasion – from hordes of eager birders wanting to see this beautiful bird.

Black Crows

Crows are a bit of a paradox. Few birds are so hated, yet few are so intelligent and fascinating. Why they are so loathed by so many people is a combination of the way they behave and their appearance: there is a long tradition of black birds being associated with evil, and birds don't come much blacker than some of our crows.

There are four black kinds of crow that you are likely to encounter in Britain: the carrion crow, rook, jackdaw and raven. Then there are two other mainly black species: the hooded crow and chough, though these tend to be confined to more remote northerly and westerly parts of the country.

Of all our birds, perhaps the crows cause the most confusion about their identity. Indeed, for two species, the rook and the carrion crow, it is said that the best way to tell them apart is whether they are in a flock or solitary: a lone bird is supposed to be a crow, whereas a flock of sociable birds is meant to be rooks. Though it is certainly true that the rook is more gregarious than its cousin, and is often found in large feeding flocks on farmland, it is also the case that carrion crows may form large flocks, especially as they go to roost at night. The confusion is made worse by the term 'scarecrow': scarecrows were originally made to scare off marauding flocks of rooks, not crows!

So if behaviour isn't a reliable way to tell them apart, how can we identify them? Well, if you get a reasonable look at a 'black bird', you should pay most attention to its overall colouring, especially on the head. Carrion crows are all black, including the bill, giving them a rather sinister appearance, while rooks have a greyish-white bill and face. Rooks are also more slender, with a longer neck and smaller head, and their bills are more pointed. In flight, these two can also

Telling apart the rook (above), carrion crow (below) and raven (below right) takes practice, but given good views is reasonably straightforward.

be told apart – the rook's style is floppier, and shows longer, more fingered wings – but it takes a lot of experience to make the distinction.

The rook and carrion crow were once the only two large crows in most of Britain, but now they have been joined in many areas by a third species, the raven. Once persecuted so much that it was seen only in remote parts of the northern and western uplands, the raven has recently spread south and east, and can now be found nesting in Kent. Ravens are huge, bulky, black birds with long wings, a large head, thick neck and massive bill, giving them a very front-heavy appearance as they fly. But the easiest way to confirm the identity of a raven is when it calls: a tremendously deep, grunting croak that you feel as much as hear!

The fourth, mainly black, species of crow, the jackdaw, is much smaller and easier to identify. Its call as it flies overhead is a distinctive 'chack', which gave the bird its name. When you see a jackdaw on the ground, it is easy to tell it apart from its larger relatives. Small and compact, it has a short, stubby bill and a very noticeable light grey patch on the back of its neck.

Our remaining two kinds of 'black' crow are not widespread. The hooded crow was once thought to be the same species as the carrion crow, but it is actually easy to tell apart, as it has large areas of grey contrasting with the black of its plumage. The 'hoodie', as it is often known, replaces the carrion crow in north and west Scotland, Ireland and the Isle of Man. The other species, the chough, has an even smaller range. It can be found only on the Scottish islands of Islay and Jura, in west Wales, western Ireland and, for the past few years, Cornwall. The chough's return to this English county in the early twenty-first century was cause for great celebration, as the chough is one of three local symbols (the other two being fishing and tin-mining) on the Cornish coat of arms. The name often puzzles people, but, of course, it used to be pronounced 'chow', rather than 'chuff', and is a representation of the bird's echoing call.

The jackdaw (left) is a sociable bird, often gathering in large, noisy flocks.

Hooded crow (below) and chough (bottom) are two of the rarer British members of the crow family.

Magpie & Jay

Of all our garden birds, surely none is so controversial as the two most striking members of the crow family: the magpie and the jay. The magpie is the pantomime villain of our garden birds, seen as a thief who takes eggs and chicks from nests, and also steals shiny trinkets and jewellery. With the possible exception of the sparrowhawk, no other bird comes close in the league table of Britain's most hated birds. And its reputation spreads far and wide – there is a famous Italian opera titled *The Thieving Magpie*.

Now that the widespread persecution of all birds is a thing of the past, the magpie population has reasserted itself. The species' habit of travelling in large, noisy flocks doesn't help its reputation either. Magpies are not subtle birds, and so we perhaps notice their presence more than we would other predators.

Magpies have adapted very well to living in our gardens. By encouraging smaller birds to feed and nest, we have created a ready supply of food for them and their hungry chicks. And this is surely the point. Just like any other creature, magpies must feed themselves and their offspring. After all, we don't condemn blue tits for killing and eating thousands of caterpillars, do we?

What is important to realise is that the population of any predator goes up and down depending on the availability of its prey, and not the other way around. Magpies are in no way responsible for the declines of some of our songbird species. Indeed, most garden birds are actually on the increase. It is birds of the wider countryside that are in decline. So why not take a closer look at a magpie, forget any prejudices you might have, and just appreciate what a truly beautiful creature it is? Superficially 'black and white', the magpie's plumage is in fact a subtle mixture

According to the well-known rhyme, a lone magpie is a sign of sorrow.

of deep greens and blues, and the iridescent feathers are set off by those patches of brilliant white. Magpies are also fascinating to watch, especially early in the year when they collect twigs with which to make their untidy and often obvious nest in the fork of a bare tree. And try to recall that childhood rhyme, 'One for sorrow, two for joy...', which was borrowed by the makers of the celebrated children's TV series *Magpie*, and which has been sung by generations of British children.

Jays are even more beautiful than their black and white cousins, but, being much shyer birds, they are not quite as familiar. One Victorian birdwatcher and writer described the jay as 'Britain's bird-of-paradise', and Bill Oddie always claims that if someone says they have seen a really unusual bird in their garden, nine times out of ten it is a jay. That pinkish-brown plumage, set off with black and white and a small patch of blue on the wings, those black markings on the side of the face, and the jaunty crest, make the jay one of our most attractive garden birds. However, it must be said that it has the same habit of taking eggs and chicks of songbirds as the magpie, but as it is a lot subtler about doing so, we don't usually see it happening.

Jays are originally woodland birds but have learned to thrive in our towns and cities, especially those with large, mature trees where the birds can make their nests. Numbers are boosted every autumn with invading jays from continental Europe, which in some 'invasion years' may arrive in huge, noisy flocks. Listen out for the jay's 'tearing muslin' call as it competes with the machine-gun rattle of the magpie – the winter soundtrack in our gardens and suburbs.

Seen well, the jay is one of our most handsome and colourful birds.

Magpies are opportunistic feeders, seeking out food wherever they go.

Woodpeckers

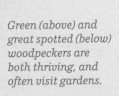

Britain has only three species of woodpecker, which are really quite different from each other. Most obviously, they differ in size: the green woodpecker is the size of a pigeon; the great spotted, the size of a large blackbird; and the tiny lesser spotted is barely the size of a great tit (and therefore very difficult to see!). Woodpeckers are more commonly heard than seen, with the characteristic drumming of the male most often experienced in woods in late winter and early spring. The drumming is either a signal to rival males and potential mates, or the sound made while excavating a nest hole. Woodpeckers seem to make rather less noise when feeding. Headbanging at up to 40 beats per second is obviously hard work, so woodpeckers have several special adaptations to cope with this behaviour. They have very strong neck muscles, and soft, spongy tissue around the base of their bill, which absorbs much of the force created by drumming. In addition, they have very sharp claws on their feet, and stiffened tail feathers, which enable them to climb up and down tree trunks, as well as hold on tightly when drumming. Woodpeckers also have very long, needle-like tongues, so long that the tongue is actually coiled up inside the skull. The tip of the tongue is also barbed, so is the ideal tool for getting at insect grubs buried deep in trees.

Green (above) and great spotted (below) woodpeckers are both thriving, and often visit gardens.

The green woodpecker is quite distinctive, both in size and colouring. They are usually first spotted sitting in the middle of a large grassy area. When they fly away, they flash their bright yellow rump and make their distinct laughing call – a 'yaffle'. The sound of all woodpeckers is traditionally meant to signify the coming of rain, hence the name 'rainbird'.

Telling the two species of spotted woodpecker apart can be trickier. The key difference is, of course, size, but

recalling the older names for these species can also help identification. The great spotted, whose plumage shows large, contrasting patches of black and white, used to be known as the 'pied woodpecker', while the lesser spotted was called the 'barred woodpecker', as its black and white markings are less distinctive. And remember that two other tree-climbing species, the blue-grey nuthatch and the browner treecreeper, can be confused with the lesser spotted, especially if you get only brief views.

The best way to find the two black and white woodpeckers is to listen for their calls or drumming, especially in late winter and early spring. As well as drumming, the great spotted also makes a distinctive, metallic 'chip' sound. If you do hear a woodpecker, try to pinpoint the direction of the sound, and then scan up tree trunks and along branches with binoculars to find the bird itself. Early in the year, when there are fewer leaves on the trees to obscure your view, is the best time to look. In winter, there is a chance of finding a lesser spotted woodpecker by looking closely at roving flocks of tits. As they roam a wood looking for food, lesser spotted woodpeckers sometimes tag along behind them.

Our three woodpeckers have experienced very different fortunes over the past few decades. While numbers of great spotted and green woodpeckers are on the rise, the lesser spotted population has been in freefall. Fifty years ago, they were common and widespread in England and Wales, but today there are just 2,000 pairs in the UK, mostly in ancient woodlands in the south.

Meanwhile, its larger relative, the great spotted woodpecker, has taken advantage of our generosity by learning to come to seed and peanut feeders and bird tables. Here, it dominates the smaller birds, which will usually flee as it approaches. Nesting blue and great tits have good reason to fear this pied predator, as they will raid nest boxes to seize their chicks.

Britain used to have another woodpecker species – the wryneck – but this has become extinct as a breeding bird and now breeds only in continental Europe. Of the ten woodpecker species that occur in Europe, just four managed to cross the Channel and recolonise Britain since they were driven out by the last Ice Age. Amazingly, Europe's largest species, the black woodpecker, is found as close to the UK as Calais but has never made the short flight over to our shores.

The scarce and elusive lesser spotted woodpecker is one of our most rapidly declining woodland birds.

Nuthatch & Treecreeper

Treecreepers sing their high-pitched, delicate song from early spring onwards.

These two characteristic birds of woods and forests are easy to overlook, yet fascinating in their habits. Apart from the woodpeckers, no other pair of birds has adapted so well to their tree-loving lifestyle.

Finding either of them can take time and patience, but it is certainly better to sit and wait, rather than walk around, as you are searching for movement, which is always much easier to notice if you are still. Listening is also helpful. Both species have distinct calls: the nuthatch has a rather loud and penetrating 'too-it', while the treecreeper utters a very high-pitched call, which can be confused with that of another small woodland bird, the goldcrest.

If you get a good view of a nuthatch, you simply cannot mistake it for any other bird. It is about the size of a great tit, but plumper and more potbellied in shape, with gunmetal-blue upper parts and orange under parts, a long, dagger-like bill and a really distinctive black 'highwayman's mask'. It also has the unique skill among British birds of being able to climb down a tree trunk as well as up, which is very useful when you live in a world of vertical trunks and horizontal branches.

The treecreeper is a smaller bird, much more modest in its appearance and habits. Basically, it is brown above and white below, but you are much more likely to identify it by its habits than by its plumage. It behaves rather like a small rodent, climbing up and around the trunks of trees before flying off to the next one, thus revealing that it is a bird and not a small mammal. If you get good views, you will see its thin, decurved bill – perfect for prising tiny insects out of the crevices of the bark in which they may be hiding.

Treecreepers are found throughout Britain and Ireland, whereas the nuthatch is confined mainly to England and Wales, though a few have now spread northwards to breed in southern Scotland. With climate change, nuthatches may continue to extend their range northwards in the coming decades, but they are a very sedentary bird, unable to cross large stretches of water, which explains why they are not found in Ireland.

Like most other woodland birds, the nuthatch and the treecreeper nest in holes or crevices in trees, but whereas the nuthatch usually chooses an old woodpecker hole, which it often makes smaller by patching it up with mud, the treecreeper prefers to nest in a narrow crack or even beneath a piece of loose bark. Both will readily take to nest boxes, the nuthatch in the usual 'tit box', and the treecreeper in a specially designed, wedge-shaped version, rather like a bat box in shape.

In the autumn and winter months, they will often join forces with other small birds, such as flocks of tits and goldcrests; tagging along with these birds is the best way to find scarce resources of food. Nuthatches, as befits their more confident character, will also come to bird feeders, often scaring off other birds as they do so.

In winter, treecreepers can be very vulnerable. They suffer especially badly during 'glazed frosts', where a freeze occurs after a spell of rain, as the frost covers up their food supply beneath the bark of trees. Numbers often drop heavily following such weather, though within two or three years the population usually bounces back.

With its colourful plumage and striking black mask, the nuthatch is one of the easiest woodland birds to identify.

Sparrows

Three birds in Britain share the name 'sparrow', two of which are true sparrows: the house and the tree sparrows. The third, the so-called 'hedge sparrow', is now more properly known as the dunnock. A member of the accentor family, it is completely unrelated to the seed-eating sparrows. Despite the name change, however, many people still refer to this charming little bird as a hedge sparrow.

Of all the birds that are seen in our gardens, towns and cities, surely the most familiar is the house sparrow. Sparrows have almost certainly lived alongside human beings longer than any other species of wild bird. They have done so from early times for one simple reason: our ancestors grew grain, which the sparrows could steal for food. We soon got our own back: communal nests of house sparrows, in the walls and roofs of our homes, were regularly raided for eggs to supplement a meagre diet.

As they have lived alongside us for so long, we have tended to take house sparrows more or less for granted. Their appearance doesn't do them any favours. They are the classic 'little brown job', though the male, at least, can claim to be a little more handsome than his mate, with his grey and brown cap and smart black bib. Female and juvenile house sparrows really are the archetypal small brown bird, with few distinctive markings, apart from a pale stripe running behind the eye.

In recent years, many of us have taken to giving sparrows a helping hand by putting up communal nest boxes. The house sparrow is a sociable bird, which prefers to nest in colonies, so a box with several entrance holes and chambers is likely to encourage them. And they need this help. In the past few decades, the sparrow has declined faster and more seriously than almost any other garden bird. This is despite the fact that we now feed wild birds much more than previous generations, which should help one so tied to human habitation as the house sparrow.

The house sparrow (right) and the dunnock (above) are often ignored in favour of more colourful garden birds, but show some fascinating behaviour.

The reasons for its rapid decline are complex. It is almost certainly down to a combination of factors, each of which may affect different populations of sparrows, and other birds, at different stages of their lives. We know that changes in farming practices – especially the sowing of winter wheat in autumn, which removes the amount of seed left on the land in winter – have affected sparrows as well as finches and buntings, their close relatives.

The 'yuppification' of our towns and cities, with loft conversions and the general tidying-up of homes, has undoubtedly reduced the number of places available for sparrows to nest under the eaves of houses. But the biggest problem may be an unseen one. It is thought that the change from leaded to unleaded petrol may have introduced a chemical into the atmosphere of our cities that kills off the small caterpillars on which baby sparrows are fed during the first two or three days of life.

If anything, the house sparrow's country cousin, the tree sparrow, has fared even worse during the past few decades. Populations as a whole have declined by almost 90 per cent, which means the species has disappeared from many places where it used to thrive. Along with other seed-eating farmland birds, such as the linnet and yellowhammer, it has undoubtedly been hit hard by modern farming methods, which reduce the amount of available food and places for the birds to nest.

If you are lucky enough to find a flock of tree sparrows, they will all appear to be males. That is because the sexes are similar. Both male and female are a shade smaller than the house sparrow, with an all-brown cap, a brighter overall plumage, and a distinct black spot just behind the ear.

The dunnock is superficially similar to a sparrow, but if you take a closer look, you will notice its more horizontal posture, slender bill, and neat, plum-and-chestnut-coloured plumage. Whereas sparrows are usually seen in flocks, the dunnock is a solitary bird, hopping about beneath the bird table and minding its own business. In spring, though, dunnocks are transformed into sexual predators. Both males and females are highly promiscuous, seeking out other partners, despite being paired up with their original mate.

The tree sparrow is one of our fastest-declining woodland and farmland birds.

Wagtails

If there's a bird walking purposefully, if a little erratically, across your closely cropped lawn, picking up insects and pumping its tail up and down as it does so, there's a pretty good chance it is a pied wagtail. The commonest of our three wagtail species, the pied is also the easiest to identify: no other British breeding bird has the combination of black and white plumage, slender shape and long tail.

Yet, despite its elegant appearance and endearing habits, the pied wagtail is often overlooked. Perhaps this is because it does not join the tits, finches and sparrows squabbling on the bird table or seed feeders. Instead, it wanders quietly but efficiently around short grass or pavements, using its sharp bill to grab the tiniest insects and other invertebrates that hide away between blades of grass or paving stones – an ecological niche it appears to have taken for itself alone.

Male and female pied wagtails do have different plumages, though you may need a close look to be certain of which one you are looking at. Males have a dark, almost black, back, and a black bib and throat contrasting with snow-white cheeks. The female also has white cheeks and a black bib, but her back is greyer. Youngsters have a less contrasting plumage, with a yellowish tinge to the head and face, giving them a rather dingy appearance, as if they forgot to wash.

The other two kinds of wagtail found in Britain are often confused with each other. Both have varying amounts of yellow in their plumage, but while one is, appropriately, called the yellow wagtail, the other, equally attractive bird is saddled with the rather misleading name of grey wagtail. So people often claim to have seen 'yellow wagtails' in the middle of winter, when this species has already migrated to Africa, and what they are actually seeing is a grey wagtail sporting a lemon-yellow plumage.

Confused? Well, the name 'grey' isn't entirely wrong: grey wagtails do have a grey head and upper parts, while the yellow wagtail is olive-green above. Yellow wagtails

The grey wagtail (above) tends to prefer more watery habitats than its cousin the pied wagtail (below).

are also much more yellow overall, with the colour extending from the face and throat all the way down the under parts, whereas the yellow on a grey wagtail is confined to the breast and belly. Female grey wagtails, and males in winter, have even less yellow on them: just a small patch underneath the tail.

The two differ in their chosen habitat as well. Grey wagtails are birds of fast-flowing rivers and streams. Like the dipper, they perch on rocks on the bank or in midstream, bobbing up and down before flying into the air to seize an unsuspecting fly. They usually build their nest in a small crack or crevice in the stone beneath a bridge. In winter, they will venture farther afield, sometimes turning up in unexpected places such as shopping-centre car parks, where, like their cousin the pied wagtail, they can find food and warmth.

The yellow wagtail is a summer visitor to Britain, breeding on wet meadows.

Yellow wagtails also like water, but of a more sedate kind: they breed in wet meadows, often alongside cattle, whose dung attracts plenty of insect food. Since World War II, much of this precious habitat has been destroyed by being ploughed up for intensive arable or livestock farming. As a result, the yellow wagtail is a much less common sight than it used to be.

Unlike its two relatives, the yellow wagtail migrates south after breeding, heading across the Bay of Biscay and Spain to Africa, where it spends the winter south of the Sahara among the big game of the African plains. In spring, it returns by a slightly different route, crossing the Sahara in a single hop in just three days, before arriving safely back in southern Britain by the middle of April.

During the winter months, the pied wagtail must often cope with very low temperatures and shortages of food. It increases its chances of survival by gathering in large, noisy roosts, often in very light places such as shopping centres or industrial estates, where it can be warm and safe from predators such as tawny owls.

Wren

Britain's third-smallest bird, after the even tinier goldcrest and firecrest, is also, perhaps surprisingly, our commonest breeding species. With upwards of eight million pairs, it is comfortably ahead of its nearest rivals, the chaffinch and robin (six million pairs) and blackbird (five million pairs), and much commoner than far more familiar garden birds such as the starling or house sparrow.

So if wrens are so common, how come we hardly ever see one? Their lack of visibility is mainly down to their shy and skulking habits. Unlike other garden birds, wrens prefer to shun the limelight, rarely venturing out into the open. They are far more likely to be glimpsed as they root around at the base of a shrubbery, or potter about a rockery, in both cases on the lookout for tiny insects, which they can grab with that short but sharp and pointed bill. It takes skill to notice a wren, and patience to get more than a brief glimpse, but if you do put in the effort, it is definitely worth it, for the wren is one of our most attractive breeding birds.

Britain's commonest bird, the wren, is also one of our smallest, weighing in at under 10 grams.

Its main feature is definitely its distinctive shape. Wrens are short and plump, with a cocked tail, which it holds up at a 45-degree angle from its body, and with short legs and a really subtle but beautiful plumage. Shades of brown, buff and black combine to give an overall chocolate-brown appearance. In flight, it whirrs along as if powered by clockwork, its tiny wings simply a blur. That small size conceals a hefty build, though, as a wren may weigh as much as 10 grams (⅓oz), about twice that of the slender goldcrest.

As with so many small and elusive birds, by far the best way to discover wrens is by listening for their sound. The male wren utters the most extraordinary song for a bird so small: a series of very loud notes and phrases, gathering speed and usually featuring a trill, once described as 'like an opera singer giving her all at the end of the aria'. No other small bird sings quite so loudly! They also have a distinctive, metallic 'ticking' call.

Wrens sing mainly in spring, often from a prominent position such as a fence post or the top of a shrub. At this time of year, the male is also very busy, as he has to build as many as half-a-dozen different nests. Known as 'cock's nests', these are carefully inspected by the female before she chooses the best one in which to lay her clutch of five or six tiny eggs. In this way, she tests out the male's commitment to her, and also picks the nest least likely to be discovered by a passing predator.

Wrens are not only found in gardens. They have colonised a greater range of habitats than any other songbird, including woods, hedgerows, farmyards, moorland, coasts and, most amazingly for a bird with such limited powers of flight, offshore islands. The isolated populations on Scottish islands such as the Hebrides, Shetland, Fair Isle and, most notably, the remote archipelago of St Kilda, have all evolved sufficiently to be considered separate and distinct races of the species. Indeed, the St Kilda wren – darker, larger and even louder than its mainland cousin – has a good claim to be a separate species, which would make it by far Britain's rarest bird.

The wren's ability to colonise new places is a legacy of its distant past. In fact, the wren is the only originally North American species to have colonised most of Europe and Asia. Back in its ancestral home, it is known as the 'winter wren', to distinguish it from 70 other species, including the cactus, marsh, sedge and canyon wrens.

Robin

Without any doubt, the robin is Britain's favourite bird. No other species – not the cheeky blue tit, the majestic golden eagle or the stately swan – can ever come close to it in the nation's affections. This is almost certainly because the robin is not only attractive in appearance but also confiding in its habits. Tameness goes hand in hand with the robin, and it is often known as 'the gardener's friend' because it will follow you around as you dig up a flowerbed.

Of course, none of this is designed to win our affections. Robins are tame because they see an easy way to get food: as we turn over the soil, so worms and other small creatures come to the surface – easy pickings for the robin. And what about that beautiful orange-red breast? It may look attractive to us, but to rival robins it is nothing less than war paint, a flash of colour designed to ward off rivals that might take over the incumbent's territory.

If a rival should dare to intrude into a robin's space, all hell breaks loose. Robins are pugnacious little creatures and will fight – sometimes to the death – to keep their right to breed in a particular place. They need to: most robins will survive only one or two winters, which means they may get just one chance to breed and pass on their genes to future generations. So this is a life-or-death battle in more ways than one.

Yet, despite their unsocial habits, which in human terms would win them an ASBO, we still love our robins. Its place in folklore is assured, and it is by far the most frequent creature to appear on our Christmas cards – a legacy of the days when the Victorian postmen wore red uniforms and were nicknamed 'robins'.

A scientist named David Lack, in the middle years of the twentieth century, was the first to discover much of the truth about the robin. Lack had the bright idea of putting different-coloured rings on the legs of the robins in his study area, which meant he could tell individual birds apart from one another. He also conducted a number of radical experiments, such as putting a stuffed robin in another bird's territory – it was promptly and viciously attacked! Lack wrote a bestselling book, *The Life of the Robin*, and later made a famous film with the BBC, entitled *The Private Life of the Robin*, which publicised his work to an audience of millions.

Despite this, several misconceptions remain about this familiar garden bird. One is that it is only the males that have a red breast – in fact, male and female robins are identical, and it is the juvenile birds

that appear brown and speckled. Another is that robins behave the same wherever they are – yet continental European robins are a shy bird found mainly in woodlands, rather than gardens.

The robin's song is among the sweetest and most attractive of all our songbirds: a plaintive series of tuneful phrases, delivered carefully, neither fast nor slow. Understandably, most people assume that only male robins sing, as with other songbirds, yet the female robin will also sing to defend her territory, especially outside the breeding season. Both male and female robins also sing throughout the autumn and winter months – unlike other garden birds, they defend a territory outside the breeding season as well.

Robins are known for nesting in some very unusual places. As well as in shrubberies and climbing plants, they will also take to open-fronted nest boxes, teapots, toilet cisterns, overcoat pockets and even the top of tractor engines!

The robin is not only Britain's favourite bird; it also has a global legacy. All over the world – especially in places once ruled by the British such as North America, Asia and Africa – all sorts of birds with a reddish or orange breast are given the name 'robin', despite having little or no connection with our familiar bird.

A robin in snow – one of the classic images of the British winter.

Starling

Is there any bird with a more contradictory public image than the starling? Hated and vilified for its supposed 'bullying' of other smaller birds on our bird tables, the very same species is celebrated for its extraordinary evening flights during winter, when millions of birds gather together in aerobatic displays before going to roost for the night.

We may try to separate these two images in our mind, but they are just two very different aspects of the lives of this much-maligned yet fascinating bird. Even if you don't like their behaviour as they squabble around your bird table, just take a moment to have a closer look.

The starling may appear black but its plumage is in fact a subtle mixture of glossy blacks, mauves, greens and browns, which in autumn and winter is heavily spotted with white. In spring and summer, the spots mainly disappear to reveal a truly stunning glossy plumage, set off by that bright yellow, dagger-shaped bill.

In late summer, people are often puzzled when they see birds that, although the size and shape of a starling, are a dull brown in colour. These are juvenile birds, which have recently left the safety of the nest and are hanging around with their parents. They always look rather embarrassed at their drab plumage – as well they might.

Although starlings may have some pretty antisocial habits, they are also fascinating to watch as they jostle for position on a bird table, often uttering little calls as if scolding their companions. In some ways they are a bit like us: sociable, noisy and a bit messy. Perhaps that's why we view them with a degree of suspicion!

Until quite recently, winter roosts of starlings could be found across much of lowland Britain, sometimes in the middle of the countryside but more often in the heart of city centres, where the extra warmth would attract huge flocks on winter evenings. There were famous roosts in the centre of Glasgow, on Bristol Temple Meads station and in London's Leicester Square. But during the past two or three decades, numbers have fallen dramatically, and now very few roosts remain – and none in our cities. The best known of today's roosts, on the Somerset Levels, once attracted as many as seven million starlings. Today, numbers are down to one or two million, but this is still enough to create spectacular aerial displays.

It is often asked how on earth birds in these gatherings avoid crashing into one another as they fly so close and so fast. The simple answer is that they look out for movements of the birds around them,

Close-up, the starling is one of our most handsome garden birds.

and when one bird shifts position – perhaps in response to a bird of prey such as a sparrowhawk or peregrine – each adjacent bird will follow suit. This creates the most extraordinary twists and turns of the whole flock, as if it were a single organism. Only when a few birds decide to go to roost for the night in the reeds below will the rest follow suit, plummeting towards the ground like water running down a plughole.

There are two possible reasons for the decline in numbers at these winter roosts. One is that the British population has been dropping, as with so many of our countryside birds, due to modern farming methods. Another is that milder winters across continental Europe have meant that many birds stay put instead of crossing the North Sea to Britain.

Starlings have another hidden talent: the ability to mimic not just other birds, but mechanical devices too. So if you hear what sounds like a car alarm or a mobile phone, but can't work out why the noise is coming from a rooftop, the chances are it is being made by a starling!

The spectacular gatherings of millions of starlings on winter evenings have become a tourist attraction in parts of Britain.

Swift, Swallow & Martins

These four summer visitors from Africa are among our most familiar breeding birds, even though they spend less than half the year with us. As they return each spring to nest in our cities, towns and villages, it is hard to believe that these tiny creatures – each of them weighing barely 30 grams (an ounce) – have travelled all the way from Africa, a distance of up to 10,000 km (6,000 miles).

Like all migrants, these birds make these journeys across the globe to take advantage of the abundant food supply available in the long days of the northern summer – in their case, billions of flying insects, which they catch on the wing to feed themselves and their hungry broods of young.

We might justifiably wonder why they don't stay put in Africa all year round, but if they did so, they would have to compete with the numerous resident species of that vast continent. By heading north, they have a greater chance of surviving and raising a family than if they stayed where they were, despite the many hazards of such a long and arduous journey.

The first of the quartet to return to our shores is generally the sand martin, which usually arrives sometime in the second half of March, though early birds do occasionally get here in February. Being birds of watery habitats, they usually head straight for lakes and reservoirs, where they can feed and replenish their lost energy on the few flying insects that are beginning to emerge.

House martins and swallows regularly gather on telegraph wires in early autumn as they prepare to migrate south to Africa.

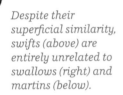

Swallows and house martins return in early April: swallows mainly to farmyards, and house martins to villages, towns and suburbs, where they build their cup-shaped nests under the eaves of our homes. Swifts – which although superficially similar are quite unrelated to swallows and martins – are the last to arrive, generally returning to our towns and cities in the last week of April and the first week of May. They are also the first to leave, mostly gone by mid-August, while the others stay until September or even October. Telling these four species apart is easier than you might think, provided you get good views. Swifts are the most distinctive, being completely sooty-black, and with narrow, scythe-shaped wings and a shortish, pointed tail. Swallows are the most graceful of the quartet, with long, swept-back wings and a long, forked tail; they are dark blue above and pale below, with a brick-red throat. The two species of martin are compact little birds with short, forked tails and triangular-shaped wings. Sand martins are brown above and white below, with a brown band across the chest, while house martins are dark blue above and white below, with a very distinctive white rump.

Despite their superficial similarity, swifts (above) are entirely unrelated to swallows (right) and martins (below).

Although these are among our most familiar summer visitors, our knowledge of them when they leave our shores and head south to Africa varies considerably. Swallows gather in huge roosts – sometimes over a million birds strong – in reed beds throughout sub-Saharan Africa. But the other three species tend to be solitary in habit, making them far harder to find in this vast continent. Over the past century or so, more than 320,000 house martins have been ringed in Britain, yet in all that time just one has been found in Africa. Soon, however, the ability to track birds by using tiny GPS devices will hopefully reveal where this familiar little bird spends the half-year when it is away from us.

The swift is the undisputed aerial master of all the world's birds. Once the chicks fledge and leave their nest, they will fly back and forth to Africa up to four times before they finally make landfall again to breed. Apart from when they are nesting, swifts spend their entire lives airborne: feeding, sleeping and even mating on the wing.

The Oakwood Trio

These three beautiful birds of western oak woodlands are among the most sought-after and entrancing of all British breeding species. There is something magical about watching them as they hop about in the sunlit branches of oak trees, flashing their contrasting colours or singing their tuneful songs. Sadly, they are nothing like as common in the rest of Britain, though migrant birds do pass through southern and eastern counties during the spring and autumn, on their way to and from their breeding grounds in Scandinavia or Scotland.

All three are long-distance migrants, spending the winter months in sub-Saharan Africa. And like so many of these global travellers, they are experiencing problems: on their wintering grounds, at their stopover places en route, and even here in Britain. In addition, climate change has led to earlier and earlier springs, which means that when these birds arrive back here in the last fortnight of April, they may already be too late to coincide their breeding cycle with the emergence of the moth caterpillars and other insects with which they feed their young. Unless they are able to change their migratory habits and bring their arrival dates forward by two or even three weeks, they may well suffer population declines as a result.

The wood warbler (above) and the redstart (below) are typical summer visitors to oak woodland in western Britain.

The redstart is arguably the most handsome of the trio. This close relative of the robin is a slim, elegant bird with a long tail. The male is especially stunning, with a black and orange-red plumage and the red tail that gives the bird its name – 'start' means tail in Anglo-Saxon. The female is drabber in colour but also sports the brick-red tail.

The male pied flycatcher may lack the bright colours of the redstart but he is equally striking in appearance. His smart black-and-white plumage isn't for our benefit, though: it is his secret weapon in the battle to win a mate. Female pied flycatchers may be rather dowdy – greyish-brown, where the male is jet-black – but they are the ones with the power in this relationship. Females check out the males not just for their song but also for the quality of their plumage – and the smartest, brightest males win the prize.

Of this western woodland trio, the wood warbler is by

far the least known. Like its close relatives among the 'leaf-warblers', the willow warbler and chiffchaff, it hides away in the foliage of the tree canopy, its delicate lemon-yellow, pale green and white plumage blending in perfectly with the new spring leaves. But during the early part of the breeding season, from late April into May, it may be more visible as it sings its sweet and memorable song: a series of echoing whistles followed by a trill, in which the bird trembles its wings until they become a blur.

The wood warbler is also the odd one out in the place it chooses to nest. Whereas the redstart and pied flycatcher are, like so many woodland birds, hole-nesters, the wood warbler builds a delicate domed cup of grass, moss and leaves on the forest floor. Of the other two species, the pied flycatcher has taken most readily to nesting in artificial boxes; one of the best-known early studies of bird-breeding behaviour was carried out by ornithologist Bruce Campbell in the Forest of Dean, monitoring pied flycatchers in nest boxes, and this research continues to this day.

The redstart, too, was the subject of one of the earliest comprehensive studies of nesting behaviour, carried out under highly unusual circumstances. During World War II, the ornithologist, poet and Oxford academic John Buxton was incarcerated for five long years in a prisoner-of-war camp in the mountains of Bavaria in southern Germany. Casting around for something to alleviate the boredom, he and his fellow POWs began to watch a pair of redstarts nesting on the edge of the camp. By the end of that summer, they had amassed a vast amount of information, based on almost constant observation of the nest. The information was later published in a charming and fascinating book, *The Redstart*, which appeared in 1950.

The female pied flycatcher, unlike the male, has a mainly brownish plumage.

Warblers

Britain's warblers fall into three main groups: the leaf warblers (willow, wood and chiffchaff), mainly found in wooded habitats; the scrub warblers (blackcap, garden warbler, Dartford warbler, whitethroat and lesser whitethroat), which prefer more open habitats; and the wetland warblers (sedge, reed, Cetti's and grasshopper), which tend to be found in or near reed beds and other wetland habitats.

The warblers are both fascinating and frustrating: fascinating, because they are one of the most diverse of all our songbird families, travelling thousands of kilometres on their migratory journeys across the globe; and frustrating, because they are often shy and elusive. Even when you manage to catch a glimpse of one, it may be hard to identify – some of our warblers really do live up to their reputation as 'little brown jobs'!

Warblers such as the willow warbler (above), blackcap (right) and whitethroat (below) migrate long distances to their wintering grounds in Africa.

And yet when you get to know them, the differences between them do begin to become clear. Some, like the blackcap and whitethroat, are actually quite easy to identify, thanks to the plumage feature that gives each species its name, but beware female blackcaps, whose crown is a chestnut-brown colour! Sedge warblers, too, are easy to identify, thanks to their streaky plumage and habit of sitting out in the open to sing.

Others, such as the garden warbler and reed warbler, are not only skulking in their habits but also more or less lacking any obvious identification feature – that is, to look at. With these birds – and with so many other warblers – you have to listen to their songs and, ideally, learn them by heart, making the birds much easier to identify. Indeed, the leaf warbler trio provided pioneering eighteenth-century naturalist Gilbert White (author of the bestselling book *The Natural History of Selborne*) with the opportunity to make his name by telling the three very similar species apart – which he did by song rather than plumage.

Cetti's and grasshopper warblers, which both live in scrubby wetland habitats, are notoriously elusive, and are almost always heard rather than seen. Fortunately, both have very distinctive songs: the grasshopper warbler does indeed sound rather like an insect, though its long-running trilling song has also been likened to a fishing reel being wound, or a bicycle wheel freewheeling. Cetti's warbler, named after the eighteenth-century Italian Jesuit priest and ornithologist Francesco Cetti, has an extraordinary song: a series of incredibly loud notes uttered in a burst of energy from deep inside the dense foliage of a bush.

Most British warblers are long-distance migrants, heading off each autumn on the long and perilous journey to sub-Saharan Africa. The exceptions are Dartford and Cetti's warblers, which are totally resident, and blackcap and chiffchaff, which are partial migrants. Most individuals of the latter two species head to Spain or North Africa for the winter, but some chiffchaffs stay put in southwest Britain, where the climate is mild enough for them to find the insect food they need to survive the cold weather.

The two resident species, Dartford and Cetti's warblers, are especially vulnerable to harsh winters. In the Big Freeze of 1962–63, the Dartford warbler was almost wiped out, but in the following decades it recovered, thanks to a run of mild winters. It can now be found in East Anglia as well as its traditional heathland home of Dorset and the New Forest. Cetti's warbler is a new arrival to our shores, having colonised southern England from across the Channel in the early 1970s. Since then, it has had its ups and downs, but, like its cousin, it is now thriving, and expanding its range farther north every year.

The blackcap's story is more complicated. Several decades ago, a small population of blackcaps from Central Europe migrated west instead of southwest, and ended up in Britain. They found a mild winter climate and plenty of food in our gardens, and prospered, so that now almost all this blackcap population winters here in Britain rather than farther south.

The chiffchaff (left) is one of the first migrants to return in spring, usually in March.

Dartford (above) and Cetti's (below) warblers are both resident in Britain, so are very vulnerable to hard winters.

Owls

No other group of birds has quite such an air of mystery surrounding it as the owl. They are both everywhere and nowhere: several species are common and widespread throughout most of lowland Britain, yet they are hardly ever seen, and even less well known. As a result, they have given rise to a large body of folklore, sayings and old wives' tales, most of which are utter nonsense, but a few do have a ring of truth about them.

Five species of owl breed in Britain: tawny, little, short-eared, long-eared and barn owls, although the barn owl is from a different family and is discussed separately. A sixth species, the snowy owl, did briefly gain a toehold as a British breeding bird during the late 1960s and 1970s, when a few birds travelled south from Scandinavia to take up residence on the Shetland island of Fetlar. A warming climate has since seen snowy owls retreat back northwards to their Arctic home, and the species is now only a very rare visitor to our shores.

Of all our owls, the tawny owl is both the commonest and most widespread in Britain, found in England and Wales and much of lowland Scotland. But even though there are around 20,000 breeding pairs (about twice as many as its nearest rival, the little owl), this species is hardly ever seen, due to its nocturnal habits.

Your best chance of catching up with a tawny owl is if you come across one at its daytime roost – you may be alerted to its presence by the noise made by small birds mobbing this predator in their midst. If you do see one, you may be surprised at how large they are, especially on the rare occasions when they take to the wing.

Tawny (above) and little (below) owls are common across much of Britain, though their habits mean they may be hard to see.

If tawny owls are hard to see, they are a lot easier to hear, though the famous 'to-whit, to-whooo' call is a myth, being a combination of sounds made by the male (hooting) and the female (the shrill 'kee-wick'). Tawny owls are our most sedentary species, with a tiny territory, so they often hoot during the autumn months – this is the male fending off any of his offspring that might otherwise take over his precious territory. They hunt by night, floating through their woodland habitat and catching voles and other small rodents.

The little owl is very different in both appearance and habits from its larger cousin. Only about the size of a starling, our smallest owl is also far more active during the day than either tawny or barn owls. You are most likely to come across one in lowland farmland in southern Britain, either perched on a barn roof or, most likely, in the branches of an oak tree. They hunt a wide range of prey, including worms, beetles and small birds and mammals.

Unlike our other four owl species, the little owl is not strictly native to Britain, having been brought here in the late nineteenth century from continental Europe, to adorn the parkland estates of stately homes. But whereas other introduced species such as the grey squirrel and ruddy duck have wrought havoc with our native wildlife, the little owl's presence here doesn't appear to cause much harm, and it is now generally regarded as a true Brit.

Two other species of owl, short-eared and long-eared, are seen much less often. Of the two, the short-eared, which often flies and hunts by day across large areas of heather moorland, is certainly the more commonly encountered. However, the male's incredible display flight, in which he rises up into the sky and then falls towards earth clapping his wings together beneath him, is a rare sight. In winter, birds from Scandinavia join our native population of short-eared owls, but, in recent years, these have become less frequent visitors.

The long-eared owl must surely take the prize for the least-known and least-sighted regular British breeding bird. It is strictly nocturnal, and frequently nests in dense coniferous forests, where birders rarely visit. Often the only clue to its presence is the strange sound made by the young birds, which sounds like a creaking hinge on a gate. In recent years, the already small population of long-eared owls has declined still further, and there must now be fears for the future survival of this mysterious and beautiful bird.

Short-eared (left) and long-eared (below) owls have very different lifestyles: short-eared hunts by day over moors; long-eared hunts by night in dense woods.

Barn Owl

Despite its obvious affinity with our other British owls, the barn owl is in fact from a separate, though related, family. It is by far the most cosmopolitan and widespread of all the world's owls, being found in six of the world's seven continents (the exception being, of course, Antarctica), and on many remote offshore islands.

Here, in Britain, the barn owl is well known as a bird of open, mainly arable farmland, where it hunts at dawn and dusk. Barn owls float over the ground on soft, virtually silent wings, thanks to a special adaptation, which makes their wing feathers almost down-like in their quality, muffling any sound. Unfortunately, this advantage is also the barn owl's Achilles heel, as its plumage gets waterlogged very easily, which means it cannot hunt when it is raining. During prolonged periods of wet weather, this can be disastrous, especially when the owls have hungry chicks in the nest. Often they will abandon their first breeding attempt and try again later in the season, if and when the weather improves.

Harsh winter weather can also be very bad for barn owls: prolonged periods with snow on the ground means that their food supply of voles and other small rodents are unavailable, and many starve to death during cold winters. The barn owl relies heavily on voles for food. Their populations rise and fall in cycles, resulting in 'good vole years', which lead to a boom in barn owls, but also 'bad vole years', when barn owl numbers may plummet, especially if they coincide with a wet summer or icy winter.

Like many predators at the top of the food chain, including the sparrowhawk and peregrine, barn owls were also badly hit by the indiscriminate use of agricultural pesticides such as DDT in the post-war farming boom. The final problem these beautiful birds face is the danger from traffic: they fly so low when hunting that they are especially vulnerable to being hit by cars or lorries, and their corpses can often be seen along major roads, especially in their stronghold of East Anglia.

With all these hazards, it might seem surprising that barn owls are actually doing reasonably well at the moment. This is partly because farmers and conservationists have joined forces to put up nest boxes in their barns and other farm buildings, which the barn owls take to readily.

Barn owls have a unique breeding strategy, to help them cope with the unpredictability of the weather and their intermittent food supply. Since they cannot predict how much food they will be able to bring back for their chicks, they begin incubating as soon as the first egg is laid, which means that the youngsters vary considerably in size and strength. In a good year, as many as six chicks may be raised to fledging, but in a bad year all but one may die. In a grisly twist to this, the oldest chicks will sometimes eat their younger siblings.

When breeding, the parent birds maximise their chances of finding food by using their extraordinarily sensitive hearing: their heart-shape face pattern and slightly asymmetrical ears enable them to focus the sound coming from an unsuspecting vole, and also to pinpoint its exact location. Once they have zeroed in on this, they drop down to the ground to grab the prey in their razor-sharp talons.

Like other owls, the barn owl has attracted its fair share of folklore. Its bizarre, screeching call, which has given it the alternative name of 'screech owl', may be uttered from spooky castle towers or battlements. This, combined with the barn owl's snow-white plumage, is surely responsible for at least some of the many ghost stories associated with these ancient locations.

The barn owl is very vulnerable to snowy weather as the snow covers up its food supply of voles and mice.

Pigeons & Doves

The monotonous but soothing sounds of cooing – whether the familiar five notes of the wood pigeon or the three notes of the collared dove – are among the classic sounds of the British summer. It is often said that the wood pigeon is giving instructions to its listener: 'Take two cows, Taf-fy...' or 'My toe is bleeding...', in each case with the emphasis on the middle syllable. The collared dove is, if anything, even more monotonous: 'I'm so bored...' is one suggested mnemonic. Others think that the bird is giving unenthusiastic support to its favourite football team: 'U-ni-ted, U-ni-ted...'.

In recent years, both these familiar birds have been doing very well, frequently appearing in the top ten of the most common and widespread garden birds. And yet they could hardly have two more different stories of success.

The wood pigeon is still essentially a farmland bird, often seen in vast numbers as they gather to feed on grain, and are then shot by irate farmers. But although still common in the countryside, in recent years the wood pigeon has moved into our towns and cities, and is now a very common sight in suburban gardens. Large and stout, it is easily told apart from other pigeons and doves by the thick white band around its neck. In flight, it has another obvious field mark: a broad white stripe down the centre of each wing.

The collared dove is another familiar fixture as a British garden bird, found in virtually the whole of the country, apart from the extreme north and west. Yet little over half a century ago, it was virtually unknown in Britain. Its extraordinary spread westwards across Europe began in the early twentieth century, and by the early 1950s it had hopped across the North Sea and gained a toehold here. The very first pair bred in Norfolk during the mid-1950s – some of today's birders recall going there specially to see them!

The wood pigeon (above) and collared dove (below) are two of our commonest garden birds, regularly coming to feed on bird tables.

Yet, twenty years later, it was a common bird in villages and towns across the country. Smaller and more pinkish-brown in colour than most other pigeons and doves, it also sports the neat black collar that gives the species its name.

The feral, town or London pigeon is regarded far less affectionately than either of these two. Its origins, however, are bizarre. Its ancestor, the rock dove, lives only on wild and rocky cliffs around the coast, and is the last bird you would have expected to colonise our urban jungles. The story began centuries ago, when our ancestors domesticated these birds for food and feathers. Later, some escaped and went feral. During the past century, others escaped or got lost when on races – despite their extraordinary navigational ability, racing pigeons don't always get home. For years, the most famous population of feral pigeons lived in London's Trafalgar Square, until visitors were banned from feeding them because of fears of disease. Today, they have another problem to contend with: peregrines, the pigeon's main predator, have moved into many of our city centres.

The feral pigeon (above) and stock dove (below left) are superficially similar, but generally live in different habitats.

A fourth resident species, the stock dove, is one of the most overlooked of all our common birds, perhaps because it resembles a smaller version of the wood pigeon, without the white collar. Close-up it is rather attractive, with a greenish-purple patch on its neck, rather like the sheen caused by a thin film of oil.

A fifth species, the turtle dove, is the odd one out, being the only British member of its family to migrate to Africa. Sadly, turtle doves are far less common than they used to be. As with so many of our farmland birds, this is partly because of modern farming methods, but it is also because thousands are shot and killed as they pass through the Mediterranean region each spring and autumn. By the way, the name 'turtle' has nothing to do with reptiles, but is a corruption of the bird's purring call, which sounds like a soft 'tur-tur, tur-tur...'.

The turtle dove (below) is one of our most rapidly declining farmland birds, due to problems at home and abroad.

Cuckoo & Nightjar

The cuckoo and the nightjar have a lot in common. Both are summer visitors here from Africa, and both are long-winged, slender birds, easily mistaken for a hawk or falcon, especially if you get only a fleeting view. They are also birds that are heard far more often than they are seen.

The call of the cuckoo is, of course, one of the best known of all Britain's birds, though the species' decline in recent years does mean that fewer and fewer people now hear them each spring. The days when observers vied with each other to report the first cuckoo in the letters pages of *The Times* are long gone. Nowadays, to hear a cuckoo call at all is something of an event.

The reasons behind the cuckoo's rapid decline are complex, but it does appear that factors at home and abroad are involved. Here, the decline of large moths means the caterpillars on which cuckoo chicks are fed by their foster parents are far less readily available. And on the cuckoo's wintering grounds in Africa, drought and habitat loss are adding to the bird's problems.

Cuckoos still arrive in Britain from mid-April onwards; the date was once commemorated around the country by 'Cuckoo Fairs' held around this time of year to welcome back the bird.

In some ways, it is rather odd that our ancestors celebrated its arrival because, as well as its distinctive call, the cuckoo is best known for a less appealing habit. Uniquely among British birds, female cuckoos lay their eggs in other birds' nests, a strategy known as 'brood parasitism'. Although this may seem rather bizarre, it clearly works: because she is free from the burden of having to raise a brood of chicks, which takes up a huge amount of time and energy, the female cuckoo is able to lay up to 20 eggs. British cuckoos have three main host species: the reed warbler, meadow pipit and dunnock. A female cuckoo will lay her eggs in the nest of the species that raised her, but perhaps the most fascinating aspect of the cuckoo's breeding behaviour is what happens after the chick hatches. Almost as soon as it is born, the baby cuckoo uses its amazing strength to throw out the other eggs in the

The nightjar's plumage is an extraordinary mixture of greys, browns and buffs, designed to camouflage the bird.

nest. This means that the unfortunate host parents have only one chick to feed – but what a chick! Cuckoos grow larger than the parents themselves, whose work is cut out feeding this monster in their nest.

Whereas the cuckoo is undergoing a sharp decline, its fellow migrant, the nightjar, is currently doing rather well. As a species on the northern edge of its breeding range in Britain, it may even be benefiting from climate change, which has enabled it to shift its breeding range farther north. Better management of its heathland and forest habitat has also contributed to its success.

Nevertheless, the nightjar remains a tricky bird to see because of its nocturnal habits. Fine, warm evenings are best: just before dusk you may hear that characteristic mechanical noise known as 'churring', which, if you are lucky, will signal the appearance of a displaying bird against the darkening sky. Before the light begins to fade, look out for the prominent white patches towards the end of the nightjar's wings, which the male uses to signal to the female. One way of getting better views of this elusive bird is to wave two white hankies around in the air, though of course you will risk being mistaken for a morris dancer!

Like owls, the nightjar has given rise to a large body of folklore: the name 'goatsucker' relates to the mistaken belief that these strange birds suckled goats. Another folk-name, 'fern owl', refers to the bird's habitat of nesting among bracken.

Nightjars are almost impossible to see on the ground, as their plumage is camouflaged to blend in with their heathland habitat. But we are now learning a little more about both these species, thanks to our new-found ability to place tiny tracking devices on them before they undertake their journeys south.

The cuckoo is far more often heard than seen, its celebrated call being linked with the coming of spring.

Skylark

Some birds are inextricably linked with a particular kind of habitat: gannets with offshore islands, robins with gardens, and mallards with park ponds. But surely no bird is quite so closely tied to the place where it lives than the skylark. This is the quintessential farmland bird, spending most of its life in and around our arable fields.

And yet the skylark also lives in another, very different, habitat. Perhaps habitat is not quite the right word, for I am referring to the place the skylark has made its own: the sky. No other bird sings for quite so long, quite so persistently, and quite so high in the air. For hours on end, you can hear them, though sometimes you will struggle to see one, so high does it fly.

When your eyes finally settle on a tiny, dark speck, appearing to bounce up and down on an invisible piece of elastic, you may well doubt that it is a bird at all. Only when you watch as it drops like a stone, descending through the air on folded wings until it reaches the ground, is its identity confirmed. Should you then follow it to try to find its nest, you are unlikely to do so; skylarks are well known for landing some distance from their actual nest site, then running unseen through the long grass to reach it, thus foiling any watching predators.

Why skylarks choose to sing in the air for so long is the result of two factors: lack of competition (only the meadow and tree pipits even try to match it, and not for long), and an evolutionary arms race in which the birds that sang for longest, and most persistently, would win the attentions of the females. They then passed on their ability to their own offspring, and so the habit developed.

Recent studies have revealed another extraordinary aspect of the skylark's behaviour. When confronted by a predator such as a merlin, experienced skylarks will sing louder, as if daring the falcon to do its worst and attack them. Less experienced birds often give in to their fear, stop singing and try to escape, at which point they are chased and caught by the merlin.

The skylark's extraordinary song flight means that it has been justly celebrated in literature and music. The nineteenth-century poet George Meredith wrote his famous poem *The Lark Ascending* with the verse and rhythm mimicking the bird's flight. The poem later inspired an equally memorable piece of music by the composer Ralph Vaughan Williams. Shelley, too, celebrated the bird in his famous ode *To a Skylark* with the opening lines 'Hail to thee, blithe spirit! Bird thou never wert'.

The skylark may have a rather dull brown plumage but it has one of the most memorable songs of any British bird.

Sadly, the song that inspired these great works of art is now rarely heard across great swathes of the British countryside. More than two million pairs of skylarks – about half the total numbers – have disappeared in the past 50 years because modern farming methods are incompatible with these birds' complex needs.

Skylarks need a range of different 'mini-habitats' if they are to thrive. These include short, cropped grass, for feeding; patches of longer grass, where their nests will be safe from predators; and, most importantly of all, stubble fields packed with weed seeds, where they can feed during the winter months. Without this mosaic of habitats, numbers of skylarks are bound to continue to fall.

If we want to get our skylarks back, it seems we have no choice. We must, as a nation, make a wholesale shift away from industrial agriculture and move back to more mixed farming, which will inevitably mean more expensive food for us as consumers. It is either that or accept that this classic bird of the countryside will continue to decline, and perhaps one day even disappear.

Pheasant & Partridges

One group of birds has achieved fame not because of their beauty, or their unusual habits, but simply because they are good to eat. Britain's game birds – those that are allowed to be shot for food at certain times of the year – come from a wide range of families, including ducks, such as wigeon, teal and mallard; waders (snipe and woodcock); and grouse. But the three most widespread and familiar species are those from the pheasant and partridge family: the common or ring-necked pheasant, and the grey and red-legged partridges.

These three species are bred and released for shooting across much of lowland England and parts of Scotland and Wales. Up to 20 million pheasants are released each year, which has made this species, pound for pound, the commonest bird in the British countryside. If the number lying dead by the sides of the road is anything to go by, it is surprising that any survive to meet the barrage of guns aimed at them from 1 October each year. But they do, and pheasant shooting is, as a result, a familiar rural sight and sound during the autumn and winter months.

The male grey partridge has a distinctive horseshoe-shaped patch on its belly.

This industry – which is, after all, what it is – is not without controversy. Animal rights campaigners object to what they regard as the wanton killing of a defenceless bird, while they also point to the waste: pheasant shooting kills so many birds that some inevitably go uneaten. Defenders of the practice point to the money it generates for the local economy, and to the creation of habitats such as hedgerows and wooded copses as nesting and roosting places for the birds, which also benefit other rural wildlife.

One irony is that the bird at the centre of this controversy is not even British. The pheasant family evolved in Asia, and the ancestors of our birds were brought here by the Romans, with numbers later augmented by those brought by another foreign invader, the Normans. And yet it has surely now been here long enough to be considered a True Brit.

The male pheasant is certainly one of the most beautiful of all our birds. Seen close-up, or even at

a distance, his bottle-green head, chestnut plumage and long tail make him unmistakable. Some birds also sport a white collar, while others appear almost black. Female pheasants are less colourful, with a shorter tail, but are still very attractive birds in their own right. Their mottled plumage is ideal for when they are nesting, as it enables them to keep hidden from predators.

The pheasant's close relative, the partridge, is an equally attractive bird. The native grey partridge, sometimes known as the 'English partridge', is smaller than the introduced red-legged, or 'French', partridge. The two are easy to tell apart: the red-legged has a series of bright stripes down its flanks, and a distinctive face pattern; the grey partridge has a horseshoe-shaped mark on its lower belly.

All three of these birds lay large clutches of eggs – the grey partridge has the largest average clutch of any British bird. Up to 30 eggs have been found in a single nest, though these were almost certainly laid by two different females. Once the tiny chicks hatch, they are able to walk almost immediately, and obediently follow their anxious parents around, finding insect food to eat.

Sadly, as with so many of our farmland birds, grey partridges have declined very rapidly over the past few decades. Whereas once they were among the top ten commonest rural birds, they have now disappeared from many areas. The best place to see them, and also the red-legged partridge and pheasant, is East Anglia, where the shooting industry still thrives. Although many people have moral objections to the shooting of game birds, it is probably true to say that without the constant release of birds into the countryside, and the care taken to maintain suitable habitats for them, the fortunes of all three species would be at a much lower ebb.

The red-legged partridge (above) and pheasant (opposite) are both non-native species, introduced to Britain for hunting purposes.

Grouse

Britain's four species of grouse are among our most iconic birds, each for different reasons. The red grouse is the classic shooter's bird, giving rise to the Glorious Twelfth (of August), the first day of the grouse-shooting season. As well as bringing much-needed cash to rural economies in northern England and Scotland, it also causes controversy as gamekeepers wage war on its predators such as golden eagles and hen harriers. In spite of its name, this bird is not red but a deep chestnut-brown, though it does sport bright red tufts above the eyes. In flight, it heads off fast and low on whirring wings, a useful ability when trying to avoid being shot.

It is no exaggeration to say that the red grouse has influenced our landscape more than any other bird, apart, perhaps, from the pheasant (see page 61). Whole areas of northern Britain would look very different if they were not managed for grouse shooting, which involves clearing trees and scrub by regular burning to ensure a fresh crop of heather for the young birds to eat.

The black grouse is no longer shot commercially, as it has become very rare in recent decades. Once found on areas of heath and moorland throughout Britain, even in the south and east, the black grouse is now confined to a few specially managed areas of moorland in northern England and Scotland. Males, known as blackcocks, are one of our most beautiful birds: a stunning combination of deep blue-black and snow-white, with crimson tufts above the eyes. Females, known as greyhens, are far less striking, like a greyer version of a red grouse.

Our largest grouse, the capercaillie, is also a very striking bird. Males, which are almost the size of a turkey, are a deep purplish-black in hue, with an ivory-coloured bill and long tail. Females are rather similar to female black grouse, though larger and with more orange-brown in the plumage. The capercaillie became extinct as a British breeding bird in the eighteenth century but was reintroduced from Scandinavia during Queen Victoria's reign. Since then, however, the species has struggled to survive, mainly because cold, wet summers mean that the survival rate of chicks can be very low.

Black grouse and capercaillie are two of the few British birds that breed using a lek, a system by which the males perform their elaborate courtship display in a kind of

The red grouse is Britain's commonest and most widespread member of its family, found in many parts of northern and western Britain.

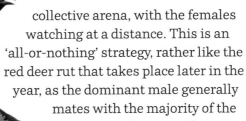

collective arena, with the females watching at a distance. This is an 'all-or-nothing' strategy, rather like the red deer rut that takes place later in the year, as the dominant male generally mates with the majority of the females. The event itself, which usually takes place at or soon after dawn, is an extraordinary spectacle for the eyes and ears, as the males leap about, uttering strange sounds, while occasionally fighting one another in a flurry of feathers.

The fourth British grouse, the ptarmigan, cannot boast quite such a complex courtship ritual, but it is unique in another way: it is the only British bird that turns completely white in winter. Like two mammals that inhabit the same mountain habitat, the stoat and the mountain hare, this serves as camouflage, helping to minimise the ptarmigan's chances of being seen and caught by predators such as golden eagles.

As spring begins to melt the snow, ptarmigans moult, at first into a patchy grey and white plumage to match the patchy snow, and later into their full breeding plumage of a greyish-brown, which in turn helps them hide among the lichen-covered boulders in their Highland home. Often the best way to locate these elusive birds is by the sound they make: an evocative croaking, which can be heard some distance away, especially on a rare windless day.

Ptarmigans are well adapted to their home on the arctic-alpine tops of the Cairngorms and Scottish Highlands. Unfortunately, the onset of global climate change may mean that they are too well adapted. As the snow begins to retreat, it is possible that we may eventually lose the ptarmigan as a British breeding bird, because they simply have nowhere else to go in this country if their habitat changes.

Black grouse (above) and ptarmigan (below) are moorland and mountain specialists, whose strongholds are in Scotland.

The capercaillie is the classic bird of the Caledonian pine forests of the Scottish highlands.

Small Falcons & Sparrowhawk

The falcons are among the sleekest and most attractive and impressive of all Britain's birds, and it is always exciting to catch sight of one. Four species breed in Britain: the kestrel, merlin, hobby and peregrine, which is discussed later, along with another unrelated bird of prey that shares their airspace and some of their habits, the sparrowhawk.

Like the peregrine, these four smaller species have all had their ups and downs in recent years. During the post-war era, when agricultural pesticides such as DDT were being used more or less indiscriminately to improve crop yields by killing insects, most British birds of prey experienced major declines. This was because the poisons became concentrated in the bodies of their prey, and the birds, being at the top of the food chain, suffered from several effects, including thinner eggshells, which reduced their breeding success.

The kestrel (above) and sparrowhawk (below) are superficially similar but have very different habits.

The one notable exception to this sad story of decline was the kestrel, which abandoned many of its former farmland haunts and took to hunting along the verges of motorways and trunk roads. As a result, it became popularly – though erroneously – known as the 'motorway hawk'. Meanwhile, the population of sparrowhawks went into freefall, while the merlin and the hobby, neither of which were common as British breeding birds, just managed to cling on, with only a hundred or so breeding pairs of each in the whole country.

Ironically, the situation has now almost gone into reverse. Sparrowhawks have bounced back, and have moved into our gardens, terrorising small birds on feeders and bird tables, with predictable results. Indeed, their predatory habits mean that sparrowhawks now rival magpies for the title of 'Britain's most hated bird', even though they are simply pursuing their natural instincts to catch food for their hungry young.

Sparrowhawks are not the only raptor success story. Merlins – which breed mainly on upland moors in Scotland and northern England – are slowly on the rise, thanks to a drop in persecution. As a result, merlins are also now a more frequent sight in autumn and

winter in southern Britain, where they often hunt flocks of finches, pipits and larks over arable fields.

The hobby has seen a far more dramatic rise in its fortunes. Once confined as a breeding bird to a few areas of southern England such as the New Forest and Salisbury Plain, in the past two or three decades it has spread right across England and Wales, and has now reached southern Scotland. As the only migrant British falcon, wintering in Africa south of the Sahara Desert, we might have expected it to suffer from problems brought about by climate change. Instead, it is thriving, and in late April and early May, flocks of more than fifty birds often gather at key wetland sites such as Stodmarsh in Kent and the Somerset Levels to hunt and feed on dragonflies. In autumn, they pass through the same areas, but this time they usually hunt young swallows and house martins, which they pursue relentlessly until they are successful.

Seen well, the hobby is reasonably easy to identify. Both sexes are dark and slender-winged, and look like a giant swift. Kestrels are also quite straightforward to identify, especially when they stop to hover – something the other three species never do. Both male and female kestrels have a brownish-orange plumage, while males have a grey head. Merlins are small, dark and short-winged, the perfect shape for chasing skylarks and meadow pipits. Sparrowhawks are very different in appearance from the others, being a true hawk. They have short, rounded wings and a long tail, ideal for manoeuvring through dense woodland foliage to pursue songbirds.

The kestrel's habit of hunting by hovering stock-still in one place, peering intently towards the ground to look for the tell-tale movement of a vole, earned it the name 'windhover', made famous by the poet Gerard Manley Hopkins in the poem of the same name. This bird's current decline may be linked to the ups and downs of the population of its main prey animal, the field vole. Whatever the reasons, the kestrel has dropped from being Britain's commonest bird of prey to the number three spot, having been overtaken by the sparrowhawk in second place and the buzzard in first.

The merlin (above) and hobby (below) are our two smallest falcons, only a shade larger than a blackbird.

Peregrine

The peregrine falcon – its name comes from a Latin word meaning 'wanderer' – is not only the fastest bird in the world, it is also the fastest creature. So it is perhaps a little surprising, and also rather wonderful, that you can now see peregrines in many of Britain's city centres.

In London and Swansea, Bristol and Bath, Derby and Cardiff, Manchester and Exeter – and many other British metropolises – peregrines are living, breeding and hunting in the airspace above the homes, offices and parks. Their main prey is the feral pigeon, though a recent study revealed that peregrines living in Derby, Bath and Exeter are also hunting by night, catching a very wide range of nocturnal migrants, including such unexpected species as Leach's petrel and corncrake!

They hunt their aerial targets by rising to a great height, then plummeting down (in a dive known as 'stooping') at up to 290 kph (180 mph) onto their unsuspecting prey. At the very last moment, just before the final impact, the peregrine puts on the brakes and grabs the bird from mid-air in its sharp talons, killing it instantly.

And yet, although pairs of peregrines regularly hunt right above the heads of millions of people, they are not always that easy to see. Oddly for such a spectacular bird, the first impressions of a peregrine in flight are not that impressive – its stocky body and broad-based wings do not always strike you as being very different from the usual gulls and pigeons. They also spend much of the time perched on a high lookout point, surveying their realm. For this reason, peregrines have been able to pass almost unnoticed by many urban dwellers.

The peregrine's natural habitat is, of course, crags and cliffs, but city tower blocks,

The peregrine is the most spectacular aerial hunter of all the world's birds, reaching speeds of up to 290 kph (180 mph).

with their high vantage points and flat ledges, are the perfect urban substitute, allowing the peregrine to live in what might at first seem to be an unnatural home.

One reason for the bird's recent arrival in our cities is that it is much more common than it used to be. During World War II, the Ministry of Defence officially persecuted peregrines, because the birds hunted and killed carrier pigeons, preventing them from taking vital messages, for example from downed airmen. At one point it was even hysterically claimed that peregrines were 'in league with Goering's *Luftwaffe*', so they were shot as enemies to the cause.

Then, as numbers struggled to recover in peacetime, peregrines were hit by another massive problem: agricultural chemicals such as DDT that thinned their eggshells and meant that most pairs failed to raise any chicks. This inspired the famous book *Silent Spring* by American author Rachel Carson, who had observed similar problems with peregrines in eastern North America. The outcry caused by the book eventually led to the banning of these chemicals on both sides of the Atlantic. Only then could the peregrine population gradually begin to recover, until now there are well over 1,500 pairs breeding in Britain.

The peregrine's new habit of nesting or perching on prominent sites in cities, such as on the roof of the Tate Modern art gallery on London's South Bank, or the Arndale Shopping Centre in Manchester, has provided a golden opportunity for the RSPB and other conservation groups to show people this amazing bird at close hand – it is the latest form of 'wildlife tourism'.

But peregrines still choose to live and hunt in their former, wilder haunts, such as mountains, moorland and rocky coasts. In winter, they often head to coastal marshes, where they will terrorise huge flocks of wintering ducks and waders, usually zeroing in on one unfortunate bird and pursuing it relentlessly, until it manages to escape or the hunter is successful. In these epic duels, time is on the side of the prey, as the peregrine is a sprinter, only able to fly at such fast speeds for a short time before it tires and must give up the chase.

Wherever we see peregrines, they are the highlight of a day's birding. They also remind us that, with careful conservation measures, birds can come back from the very brink of extinction, to thrive and delight us once again.

Buzzards & Harriers

The common buzzard (above) and honey buzzard (below) are superficially similar in appearance but lead very different lifestyles.

Like our smaller raptors, Britain's larger birds of prey also suffered during the decades following World War II. This was due not just to pesticides, but also because they were relentlessly shot, poisoned and trapped by gamekeepers and landowners who regarded any bird with a hooked beak and sharp claws as the enemy.

Fortunately, in these more enlightened times, shooting and killing are on the wane, and most of our larger birds of prey, including our two breeding species of buzzard and three harriers – are thriving. The exception is the hen harrier, which, because it lives and breeds on grouse moors in northern England and Scotland, and preys on young red grouse, is still subject to illegal persecution, keeping its breeding numbers artificially low.

Of all our large raptors, by far the commonest, most widespread and most easily seen is the common buzzard. This is one of the late twentieth century's most uplifting conservation success stories. Just 20 years ago, when the British Trust for Ornithology (BTO) carried out their second nationwide *Atlas* survey of our breeding birds, the buzzard was confined to the northern and western areas of Britain – the uplands. Indeed, in Scotland it became known as the 'tourist eagle', as so many visitors would claim to have seen a golden eagle, when in fact all they had seen was a buzzard.

In southern and eastern England – apart from a few strongholds such as Dartmoor and Exmoor – the buzzard was virtually absent. Yet now it is our commonest bird of prey, having overtaken both the sparrowhawk and kestrel to reach the number one spot. It now breeds in every English county, even as far eastwards as Norfolk and Kent. This is almost entirely due to the huge drop in the persecution of raptors.

Buzzards sometimes cause identification problems for the unwary, due mainly to their very variable shading and colours. The default buzzard is brown above and brownish below, with a few paler patches on its under parts. But some birds can be almost entirely dark brown below, while others can appear virtually white, giving rise to regular mistaken reports of 'ospreys'.

Buzzards are often seen soaring high in the sky on their broad wings, unlike the honey buzzard, one of our rarest and most secretive birds of prey. A summer visitor from Africa, honey buzzards feed not on the honey itself, but on wasp larvae, raiding the nests of these social insects to get at them. Honey buzzards have evolved the ability to fend off stings by having specially stiffened feathers around their head and face. Once incredibly rare in Britain, this species is now on the rise, and has spread into Wales from its southern English stronghold.

Two of our three harrier species also migrate to Africa, with the exception of the hen harrier, which tends to winter on coastal marshes. Female hen harriers, known as 'ringtails', are brown with a white rump; males also have the white rump but are a beautiful pale grey in colour.

The marsh harrier is our commonest harrier species, and another conservation success story, rising from just one or two pairs in the early 1970s to over 300 pairs today. It has also spread north and west from East Anglia, and now nests in southern Scotland. Male marsh harriers are distinctively patterned with contrasting grey, black and brown on the wings; the much larger females are dark chocolate-brown with a creamy-yellow cap.

Our rarest harrier – and indeed Britain's rarest breeding bird of prey – is the slender and elegant Montagu's. Named after the pioneering eighteenth-century ornithologist George Montagu, who first realised that this and the very similar hen harrier were different species, it nests on farmland in eastern and southern England, though its breeding sites are often kept secret to avoid being targeted by egg collectors. If you are lucky enough to see a Montagu's harrier, it may strike you as looking more like a falcon than a harrier, as it bounds through the air on its stiff yet buoyant wings.

Our three species of harrier, hen (left), marsh (above) and Montagu's (below), have experienced very different fortunes in recent decades.

Eagles

Our two largest birds of prey are also our most sought-after, living as they do in remote and sometimes inaccessible regions of Scotland. Searching for them can be frustrating, especially as buzzards and hen harriers may also be present where they are found. But when you finally see the real thing, you can be in no doubt: the golden eagle is a huge, majestic bird, with long wings and a distinctive rectangular silhouette. The white-tailed eagle – also known as the sea eagle – is even larger, and has memorably been described as a 'flying barn door'.

The golden eagle is perhaps the iconic bird of Scotland – only the osprey rivals it for fame. It is a very large raptor, with a wingspan of over 2 m (6½ ft), and is usually seen soaring majestically over its mountain habitat on its long, broad wings. If you are lucky enough to see one close-up – a rare sight indeed – you may be able to spot the patch of golden feathering on the sides of its neck that gives the species its name. There are also often patches of white on the wings, rump and tail.

Golden eagles are quintessentially birds of the high mountain ranges of the Scottish Highlands, though they do occasionally venture down to sea level in parts of the western coast and islands. They hunt a wide range of prey, from mountain hare and ptarmigan, on the high tops of the Cairngorms, to wildfowl, red grouse, deer and sheep, though these larger mammals are generally taken as carrion. Golden eagles do not breed quite as high as we often suppose, mainly because it is easier – and less energy-consuming – to carry heavy prey downhill rather than uphill!

Scottish golden eagles are now doing rather well, and in recent years birds have colonised England's Lake District, and they have also been reintroduced to their former haunts in northwest Ireland.

White-tailed eagles are much more birds of the lowlands and coastal areas, as their alternative name of 'sea eagle' suggests. Indeed, they are much easier to see than their relative, often sitting out in full view on estuaries or mudflats, or flying low along the coast. Seen well, they are impossible to mistake for any other British bird: they are simply vast in size – like a large dog – and have a pale head and bright yellow bill. When they fly, they reveal the very distinctive white tail that gives the species its name, though this is lacking in young birds, which are mainly brown in colour.

Golden eagles were persecuted almost to extinction in the nineteenth and early twentieth centuries because they hunt game birds and feed on dead sheep, making them powerful enemies among the farming and shooting communities of Scotland. They were reprieved only because they live in such remote areas. Sadly, the white-tailed eagle did not manage to survive, and died out as a British breeding bird, with the last pair breeding on the Isle of Skye in 1916. For more than half a century, the white-tailed eagle was absent from Britain, until, after several unsuccessful attempts, birds were reintroduced from Scandinavia to the Isle of Rum in the mid-1970s. After a slow start, the birds thrived, and white-tailed eagles now breed right along the Scottish west coast, and have recently been reintroduced to the east coast as well.

White-tailed eagles are one of our great conservation success stories. Despite much local opposition at the time, the project not only succeeded, but now contributes hugely to the local economy in the form of wildlife tourism. So as well as visiting the Isle of Mull to see the location of the children's TV series *Balamory*, many people also come to see these wonderful birds, which have, of course, usually featured on *Springwatch* since the very first series. Who could forget the antics of the youngsters Itchy and Scratchy, named by local schoolchildren, as they grew, fledged and finally left the nest?

The white-tailed (or sea) eagle has now re-established itself as a British breeding bird, thanks to a long-running programme to reintroduce the species to Scotland.

Osprey

Ospreys are supremely adapted to prey on large fish, which they seize from the water using their sharp talons.

Despite its comparative scarcity, the osprey is arguably our best-known bird of prey. This, and indeed the fact that it breeds in Britain at all, is almost entirely due to the foresight of one man, the pioneering ornithologist and conservationist George Waterston.

It was Waterston who realised that the best way to protect a pair of ospreys that had returned to nest at Loch Garten in Speyside was to publicise the event and invite the Great British Public to come along and enjoy the birds. Over the years, they came in their millions, and today the RSPB's reserve at Loch Garten is one of the best-known birding sites in Britain.

But it is another site that has produced Britain's most famous osprey – indeed one of Britain's best-known individual birds. Loch of Lowes, in Perthshire, is home to a long-lived female osprey known as 'Lady', whose dedication to rearing generations of Scottish ospreys has been celebrated on TV, in newspapers and in a recent book.

The strange thing about ospreys is that although they are so cherished here in Britain, across the Atlantic in the United States they are dismissed as 'trash birds'. This is perhaps because they are so common and widespread, especially along the east coast, where they build their huge nests on platforms in busy harbours and marinas. They are also very common throughout both the Old and New Worlds, being found on every continent apart from Antarctica.

Wherever ospreys live, they feed on a specialised diet of fish, which they catch in a truly spectacular fashion. Flying low across the surface of a lake – or, where they can find it, an artificial fishpond – they dip their talons down at the last moment to grab an unsuspecting fish such as a salmon or trout. Their claws are especially adapted to grip the slippery fish so that it does not fall from their grasp.

As well as this unique habit, ospreys are easy to identify by their distinctive appearance. About the size of a buzzard, they are our palest bird of prey, being almost all white beneath, and brown above, with a white head crossed by a brown stripe through the eye, giving it a bandit-like appearance, and a distinctive loose crest. In flight, they appear almost gull-like, as their wings are much floppier than other birds of prey, and kinked in the middle.

Ospreys are often seen in southern Britain as they migrate to and from their winter quarters in Africa, sometimes stopping off to feed at places such as the Somerset Levels or the London reservoirs. We can now track their travels thanks to the dedication of another great ornithologist, Roy Dennis, who puts tiny satellite tracking devices on young ospreys before they head off on their epic global journeys.

In recent years, ospreys have also been reintroduced into central England at Rutland Water. Here, as at Loch Garten, they attract a steady stream of visitors keen to see this iconic bird. Birds from this population may also have been involved in the recent colonisation of Wales, where ospreys now also breed regularly.

And yet it all could have been so very different. Persecuted for centuries, like so many of Britain's raptors, ospreys finally disappeared from Scotland in the early twentieth century. When a single pair returned to breed at Loch Garten in 1954, they had to run the gauntlet of egg collectors, desperate to get their hands on a clutch from such a rare bird; these thieves foiled several breeding attempts.

But thanks to the vision of George Waterston, and the dedicated work of hundreds of volunteer wardens over the last half-century or more, ospreys are now not only common and widespread in Scotland but are also spreading south. Perhaps in years to come this once rare and sought-after bird of prey will be a common sight throughout southern Britain.

The osprey is one of the twentieth century's greatest conservation success stories: the birds now breed in England and Wales as well as Scotland.

Red Kite

The red kite is surely the most elegant and graceful of all our raptors. As it soars overhead on long, kinked wings, using its long, forked tail as a rudder to help steer its way through any change in the wind currents, it looks like a sailing ship on the oceans.

Its colour, too, is extraordinary. It really is red – a vibrant, brick-red, in some lights almost orange, contrasting with patches of black, brown and buff. You may see an awful lot of buzzards when you are searching for red kites, but, believe me, when you see the real thing, you can be in no doubt that you are looking at a red kite.

The red kite also has an extraordinary tale to tell. It is one of the greatest conservation success stories of all time, involving a comeback from the very brink of extinction as a British breeding bird, to take its rightful place in our skies once again.

Red kites weren't always so rare. Indeed, in Shakespeare's day they were one of the commonest birds in our cities, with flocks gathering in the streets of London to feed on human waste. In some ways they performed a similar role to gulls and foxes today: they were the clean-up squad. As such, they were at least tolerated by the inhabitants of Elizabethan Britain.

But soon afterwards their status – and attitudes towards them – began to change. Along with other birds of prey such as the buzzard and golden eagle, kites represented a threat to the new landowning interests of the eighteenth and nineteenth centuries. Money could be made from rearing pheasants and partridges for shooting, and so every bird with a hooked beak and sharp claws was categorised as 'vermin' – to be shot, trapped and poisoned wherever possible.

The effect on the kite population was devastating, no doubt made worse by these birds' confiding behaviour and their habit of flying low, which made them easy targets. By the end of Queen Victoria's reign, at the turn of the twentieth century, the UK's red kite population was

Red kites are our most graceful and buoyant bird of prey, as they soar using their long wings and forked tail to keep airborne.

down to a tiny handful of birds, confined to their last sanctuary, the wooded valleys of mid-Wales.

And there they remained well into the post-World War II era. Until the late 1980s, the only place you could reasonably expect to see red kites in Britain was in those same hidden valleys. The bird had even become a symbol of Welsh resistance against the English. Even here, though, red kites were far from safe. Their extreme rarity had made them a target for a small but determined band of egg collectors. Only the efforts of the RSPB, helped by volunteer guards, including, on occasion, soldiers from the British Army, allowed the kites to maintain a precarious toehold as a British breeding bird, and avoid the fate of the osprey and white-tailed eagle, both of which had become extinct in Britain.

The tide finally began to turn in the late 1980s, when a brave and far-sighted decision was made to reintroduce the red kite into parts of its former range in England and Scotland. Initially, birds came from Scandinavia, but, being migrants, they soon flew away. Kite chicks were then brought from Spain, and, after a few teething troubles, they managed to re-establish viable breeding populations in the Chilterns to the northwest of London, and in the East Midlands near Stamford. Two other projects, in Dumfries and Galloway, and Black Isle in Scotland, fared less well, mainly because local landowners still viewed the birds as a threat and they were often poisoned.

Soon red kites became a familiar sight, especially over the M40 motorway near Oxford, and locals even began to feed them bits of meat and chicken bones, making them a regular, if unlikely, garden bird. At the same time, the Welsh population was also thriving, partly thanks to a change in the attitudes of farmers and other landowners, who began to realise that the kite was a potentially lucrative tourist attraction. One particular farmer, at Gigrin Farm in the kite's mid-Wales stronghold, began to feed them, attracting thousands of visitors keen to see this spectacular bird in action.

And now, at the start of a new millennium, red kites are on the verge of returning to London itself. They may not be so bold as to feed in the streets again, but we can look forward to seeing this magnificent raptor in all its glory soaring over the capital once more.

Auks

To live the vast majority of your life at sea requires a whole range of adaptations to deal with one of the most alien and, at times, hostile environments on earth. The group of seabirds known as auks has evolved to be supremely well suited to a life on the ocean wave – sometimes literally, as they spend much of the year floating on the water's surface.

Watching an auk in flight, frantically flapping its short, stubby wings in order to stay airborne, you might wonder why their wings aren't longer, but watch the same bird underwater and it is transformed. The wings act as powerful flippers, driving it rapidly through the water and enabling it to twist and turn in an instant to chase and catch fish. Indeed, when underwater, auks resemble another group of ocean-going birds, the penguins, though, of course, they have not yet dispensed with the ability to fly.

The most familiar British auk is undoubtedly the puffin (see page 84), but there are three other species that breed around our coasts: the guillemot, razorbill and black guillemot. Of these, by far the most common and widespread is the guillemot. These birds breed in vast colonies, lining the cliff-ledges and joining in with the cacophony of sound that so characterises our famous seabird spectacles. Although you can see guillemots from any coastal headland, they are far more prevalent on the sea-cliffs and islands of the north and west, especially Pembrokeshire in southwest Wales, the Farne Islands off Northumberland, and almost anywhere around the Scottish coast.

From a distance, guillemots appear basically black and white. Seen closely, they are actually chocolate-brown above and white below, with a slim, dagger-shaped bill tapering to a point. You may notice that some birds have a neat, white eye-ring and white mark extending across their face – these are known as 'bridled' birds, which become more common the farther north you travel.

Living cheek-by-jowl with its neighbours, the guillemot has the smallest territory of any breeding bird – just a square metre or so. The reason for this is that they need to avoid predators such as foxes and rats, and, since their food is widely available out at sea, it makes more sense to gather together in places of safety than to try to defend their food supply.

The guillemot is the commonest and most widespread of Britain's auks, nesting on sea-cliffs around the British coasts.

Guillemots lay a single, pear-shaped egg on a narrow cliff-ledge. Famously, if this egg is knocked, it rolls in a tight circle, barely shifting from its original position – the shape of the egg has evolved over time to ensure it doesn't fall off the narrow ledge where the birds breed. Less than two weeks after it hatches, the chick – still not fledged and wearing its original downy plumage – launches itself bravely into the air and plummets to the sea below. This may look like an act of madness, but if the growing chick stayed out on the cliff-ledge, it would become increasingly vulnerable to attack by gulls.

The guillemot's close relative, the razorbill, is darker and chunkier in shape, appearing almost black above and snow-white below. Its most obvious feature is the distinctive thick-set bill that resembles an old-fashioned cut-throat razor, which gives the bird its name.

Both razorbills and guillemots hunt for fish in deep water, diving from the surface to depths of up to 90 m (300 ft), equivalent to the height of Big Ben.

The third member of the auk trio, the black guillemot, is less familiar than its larger and commoner relatives. It is mainly found in the north and west, and can often be seen swimming close to the shore in harbours, especially in the Western Isles, Orkney and Shetland. The black guillemot is also known as the 'tystie' – a Shetland name deriving from its soft call. It is an attractive little bird, smaller than a guillemot, and all dark brown in summer with a white patch on its wings. Its bright crimson bill and feet give it a comical appearance.

There used to be another auk species breeding in Britain: the great auk. Sadly, however, this giant flightless version of the razorbill was hunted to extinction by the middle of the nineteenth century. It is to be hoped that the current rapid decline of fish stocks will not lead to the same fate for any of its smaller relatives.

The razorbill (left) and black guillemot (below) are common around much of the northern and western coasts of Britain.

Underwater, the guillemot reveals its ability to dive to extraordinary depths to catch fish.

Puffin

The puffin is something of a paradox. It rivals the robin and the blackbird as one of our best-known and best-loved birds, yet most people have never actually seen one in the flesh! We may be surrounded by puffin soft toys, puffin cartoon characters and even Puffin Books, but the bird itself is far more elusive, being confined to remote headlands and offshore islands around our coasts.

The puffin is a small, plump, mainly black and white seabird, belonging to the auk family. What separates it from its closest relatives, and has led to the huge affection we feel for it, is without doubt its extraordinary bill. No other British bird, on land or sea, has quite such a colourful appendage: a kaleidoscope of reds and yellows, giving it the appearance of a tropical creature. Indeed, before it gained the name we use today, the puffin was called the 'sea parrot' until at least the eighteenth century.

Sand eels are the puffin's favourite food, but shortages are causing problems for many puffin colonies.

Puffins, like other members of the auk family, live the vast majority of their lives at sea, returning to land only for a few months during the spring and summer in order to breed and raise a family. They nest not on the ledges of sea-cliffs but in deep holes, usually old rabbit burrows, on the grassy slopes above. To avoid attack by predators such as large gulls, they fly in very fast, crash-landing a metre or so away from the burrow entrance. They then run as fast as their bright red legs can carry them, before plunging down the hole to safety. Once inside, they often emit a bizarre moaning noise, which to our ears sounds rather amusing.

Puffins usually lay a single egg, and feed their hungry offspring sand eels for about six weeks. They then desert the chick, which remains in the burrow for another few days, until hunger drives it to take the giant step into daylight and find food for itself.

Recently, however, a shortage of the puffin's staple diet of sand eels has meant that the adult birds have been forced to find other, far less suitable, food, such as pipefish – this bony creature is completely unsuitable for the chicks, as they often choke while trying to eat it. The lack of sand eels is due to a combination of over-fishing and the effects of global climate change, which are driving the sand eels farther north because of rapidly warming seas.

After the breeding season is over, the adults and their youngsters head out to the open ocean, where the adults shed the brightly coloured sheath on their bills for a smaller, drabber version. The birds remain out of sight for the rest of the autumn and winter, bobbing about on the stormy seas and diving to find food. Until recently we had very little idea of where puffins spent their time away from us, but the use of tracker devices is now beginning to solve the mystery. Scientists have discovered that most British puffins venture into the middle of the North Atlantic Ocean, far from land. They come back to their breeding colonies in March or April and settle down to breed soon afterwards, remaining there until July or early August, before they head out to sea once again.

Like all auks, the puffin breeds on remote cliffs and islands to avoid predators.

If you want to see puffins for yourself, there are a number of accessible colonies around the country. The easiest is the mainland colony of Bempton Cliffs on the Yorkshire coast, where the birds can be watched from the cliff-top. Puffins are also common on the remote offshore islands of St Kilda, to the northwest of the Outer Hebrides, and the islands of Unst and Noss on Shetland. There are also a few puffins on Lundy, off the north Devon coast – indeed, 'Lundy' means 'puffin island'.

But if you want to see puffins in large numbers, the best places to visit are the island of Skomer, off the coast of Pembrokeshire, or the Farne Islands, off Northumberland. Here you can get unbelievably close to these little birds, and have the ideal opportunity to take photographs to remind you of the experience.

Gulls

The haunting cry of gulls takes many of us back to early memories of seaside holidays, building sandcastles and swimming in the ice-cold sea. Yet this sound has become, in the past few years, increasingly common in many of our city centres, some of them far inland. For what we once thought of as 'seagulls' are moving away from their traditional homes by the coast and into our own neighbourhoods.

When they get here, they are not always very popular. Noise is one issue, but perhaps the biggest problem is that they intimidate us by hanging around like groups of lager-louts, fighting over scraps of food or, worse, snatching it from our hands.

In some ways, this is a very modern phenomenon, but gulls have, in fact, lived inland for at least a hundred years, and their population has steadily increased during that period. The first species to come regularly to our towns and cities were two smaller ones: black-headed and common gulls. They are a familiar sight on playing fields and park ponds, where they will often feast on bread thrown to the ducks. In the evenings, they gather to roost on reservoirs, where they are safe from predators such as foxes.

Black-headed gulls are the smallest members of their family commonly seen in Britain. In their breeding plumage, they sport a smart, chocolate-brown hood and red bill, but outside the nesting season they lose the hood, so that only a small dark spot remains just behind their eye. They are a slender bird, rather pigeon-like in appearance, with pointed wings. Common gulls are larger and plumper, with a green bill and legs. Despite their name, they are rather scarcer than the black-headed gull.

In the past decade or so, two much larger relatives – the herring and lesser black-backed gulls – have joined them. These mainly marine gulls have entered our city centres for two reasons: a readily available supply of food from discarded takeaways and local landfill sites, and plenty of safe places to nest, such as the flat roofs of office buildings and car parks, where they can raise their young without fear of attack by foxes.

These birds have also deserted the coasts in large numbers because fish stocks are much lower than they used to be, and changes in fishing practices mean there are far fewer fish being gutted in harbours, resulting in less food for the birds. The coastal population of herring gulls has consequently declined so rapidly that

Gull identification can be tricky – check out the leg and bill colour of the common gull (above), lesser black-backed gull (below) and herring gull (opposite above).

they are now on the Red List of birds of conservation concern.

The lesser black-backed gull has benefited in different ways from the ready availability of food on landfill sites. This species used to migrate south in autumn to spend the winter in Spain and Portugal, even West Africa. Nowadays they stay put, as there is plenty to eat all year round.

These two larger urban gulls can easily be told apart by the colour of their upper parts: the herring gull's back and wings are a very pale grey, whereas those of the lesser black-backed gull are a much darker grey – almost black at a distance. Another even larger species, the great black-backed gull, has a virtually black back. It is, however, still found primarily on coasts, and rarely ventures inland.

The other two regular British breeding gulls are less frequently seen. The kittiwake is mainly an ocean-going bird, nesting on cliffs and islands alongside guillemots and razorbills. However, a small colony also breeds on Tyne Bridge in the centre of Newcastle, almost 16 km (10 miles) from the sea! The Mediterranean gull is a recent colonist, and is now thriving in parts of southern England, especially the Kent coast and Poole Harbour in Dorset. It is a very smart gull, all white with a jet-black head and blood-red bill, and no black on the wingtips in flight, which gives it a rather ghostly appearance.

One other seabird, the fulmar, closely resembles gulls but it is in fact related to the shearwaters and petrels (see page 118). It can be seen in seabird colonies, but it has also spread rapidly in the past century and is to be found almost anywhere around our coasts. Look out for a gull-like bird hanging acrobatically in the air, but keep your distance – nesting fulmars have a nasty habit of vomiting sticky and very smelly oil on anyone who gets too close!

The kittiwake (above) and great black-backed gull (below left) are our two most marine species of gull, rarely venturing inland.

The Mediterranean gull is a recent colonist as a British breeding bird, having first nested here in 1970.

Terns

If gulls show a modicum of grace in flight, their relatives the terns beat them hollow. As one observer noted, a tern is like a gull that has died and gone to heaven. Their floating flight, pointed wings and aerobatic manoeuvres do give them a rather angelic quality, and always mark them out as something different from their run-of-the-mill relatives. Their old name, the 'sea-swallow', is appropriate, as all terns have long, pointed wings and long, forked tails.

Five species of tern regularly breed in Britain. All are summer migrants to our shores, heading off to spend the winter months in West Africa – or, in the case of that epic global traveller the arctic tern, even farther south.

The tern most likely to be seen in Britain is, appropriately, the common tern. They are not only found on coasts, but have also moved inland and often breed on artificial rafts provided for them on reservoirs and gravel pits. They can even be seen along the River Thames and canals in the middle of London. Look out for them from the middle of April, when newly arrived birds are often very vocal as they try to find a mate and settle down to breed.

The common tern is a medium-sized seabird, about the size of a small gull, with grey upper parts, white under parts, a black cap and a crimson bill tipped with black. The arctic tern is very similar in appearance, though if you get good views, you will notice it is darker, almost blue-grey, beneath, with shorter legs and a blood-red bill without the black tip. Both have a very buoyant, floating flight, and will fly low over water before plunging down to grab a fish with that sharp, pointed bill.

Arctic terns breed in large colonies off our northern and eastern coasts, especially on Shetland, the Farne Islands and north Norfolk. After the breeding season is over, both adults and youngsters undertake the longest regular global journey of any living thing, flying all the way to the opposite end of the earth to spend the winter in the seas around Antarctica. As a result, an arctic tern sees more hours of daylight in its lifetime than any other creature, and

The arctic tern undertakes the longest global journey of any living creature, migrating to and from Antarctica each year.

may fly well over a million kilometres on its epic journeys.

The Sandwich tern is the largest of its family regularly seen in Britain. It is one of the earliest migrants to return in spring, usually appearing by late March, and is named after the place where it was first shot and identified: the town of Sandwich, on the coast of east Kent. These are large and stocky terns, more like a gull in general appearance, with a shaggy black cap, white forehead and harsh, grating call.

The smallest British tern, the little tern, is truly tiny, barely longer than a song thrush. It is best identified by its size, together with its yellow and black bill and fluttering, almost moth-like, flight. Like their relatives, little terns are aerial hunters, hovering over shallow water before plummeting down to grab their prey.

Little terns are declining rapidly in Britain because of their unfortunate habit of breeding on shingle beaches. Although this enables them to camouflage their clutch of eggs, it also makes them very vulnerable to disturbance by dogs and people. To combat this, local conservationists often rope off the area where the birds breed.

The fifth member of the family to breed in Britain, and also by far the rarest, is the roseate tern. This elegant bird is so named because, during the breeding season, the birds often show a rosy-pink hue on their under parts. At a distance, they often appear pure white, with a black bill and long wings and tail. Roseate terns nest in only a few places in Britain, with the largest colony being on Coquet Island, to the south of the Farne Islands, off the coast of Northumberland. Roseate terns migrate to West Africa in early autumn, where they may be trapped and killed, so international protection measures will be needed to safeguard their future as a British bird.

Little (above) and sandwich (below) terns both migrate to Africa to spend the winter there.

Skuas

The great skua is also known by its traditional Shetland name of 'bonxie'.

Skuas represent the dark side of seabirds, chasing their smaller relatives in order to make them disgorge or drop their food, which the skua can then grab for itself. This practice – known technically as kleptoparasitism but to most of us as piracy – has enabled them to become highly successful. And there can be no doubt that watching a skua chase and harry a tern or a kittiwake until the unfortunate bird has regurgitated its catch is impressive.

Two species of skua – the great and the arctic – breed in Britain. They are closely related to gulls, and, indeed, the great skua bears a close resemblance to a young great black-backed or herring gull, being large, bulky and dark brown in colour. Unlike gulls, however, they are not very familiar to us, as both species breed only on our northernmost offshore islands and headlands, such as St Kilda, Orkney and Shetland. Although you may sometimes see them around coasts farther south, especially during onshore gales in autumn and winter, for most of the time they are usually far offshore and out of sight.

If you visit their breeding grounds at the height of summer, it's a very different story. Here you can get close to skuas, but beware! Get too near their nest, where they guard their precious eggs or chicks, and they will attack, flying fast at you, then, just before they reach their target, pushing out their feet towards your head before veering away. Being pursued by a bird the size of a great skua is no joke, as they weigh several kilograms, and can pack a hefty whack with their broad wings. Arctic skuas may be smaller, but they are faster and more manoeuvrable, so being attacked by them is a bit like coming under fire from a military jet fighter.

Of course, the birds bear you no real malice; they are simply defending their territory, eggs and chicks. But this is not much consolation when it happens. If you come under attack, the best way to avoid being hit is to hold a stick above your head, which will at least stop the birds getting too close.

Both great and arctic skuas are very handsome birds. The great skua is basically brown in colour, with lighter feathering on its head, neck and back, a thick crown and a huge dark bill. Arctic skuas are smaller and more elegant, recalling a tern in flight. Oddly, arctic skuas come in two quite distinct colour forms, known as 'light phase', when the under parts are yellowish, and 'dark phase', when they are chocolate-brown. This colour difference is nothing more than a minor genetic variation, and both types breed happily with each other.

From a global point of view, great skuas are in fact our rarest breeding seabird. Almost three-quarters of the world population breeds in Britain, half of them on Shetland. Here they are often condemned for their habit of killing other seabirds or for scavenging by the side of the road, but they show a more delicate side when bathing communally in the shallow lochans that dot this treeless landscape.

From time to time, skuas turn up in unusual locations such as inland reservoirs. This usually occurs after heavy westerly gales in autumn, when the birds may be blown inland. After feeding for a few hours, they usually regain their bearings and head back to sea, their brief stay in unfamiliar territory over. They may occasionally be joined by two rarer skua species: the long-tailed and pomarine skuas. Both breed farther north, but pass through the British Isles on their way to and from their summer homes in the far north, and winter quarters in the South Atlantic.

Arctic skuas come in two distinct colour phases, light and dark – this one is dark phase.

Ringed & Little Ringed Plovers

The plover family includes some of our most attractive wading birds, such as the golden plover and lapwing. But it also features two small waders, which are easy to overlook and often confused with one another: the ringed and little ringed plovers.

Plovers have certain adaptations, which make them different from other wading birds, most notably a short, stubby bill. This they use to pick up items of food from the ground rather than digging into the soil or mud, as do longer-billed birds such as curlew and snipe. Plovers also have short legs compared to their body size, which means they often run in a characteristic way, dashing a metre or so forward, bending to pick up a morsel of food, then running forward a few more metres. The name 'plover', incidentally, comes from the Latin *pluvialis*, meaning 'rainy', – the birds used to migrate through Europe in autumn, their arrival coinciding with rainy weather.

These two 'ringed' plovers are superficially very similar in appearance. Both are small, plump wading birds with brown upper parts, white under parts, a black band across the breast, and a black mask across the face. Look closer, however, and there are clear differences. The ringed plover has bright orange legs and an orange base to its bill, while the little ringed plover has a dark bill and greenish-brown legs. It is also a slimmer bird, with an altogether more elegant appearance. In flight, the two are easier to tell apart: the ringed plover has an obvious white wing-bar, whereas the little ringed plover's wings are plain brown. But the best field mark is that little ringed plovers have a distinctive bright yellow ring around each eye, giving them a rather dapper appearance.

The two birds also usually have different habitats. The ringed plover is primarily a coastal bird, living on estuaries and salt marshes in autumn and winter, and breeding on shingle beaches in spring and summer. Little ringed plovers are rarely seen at the coast, and are generally found inland on gravel pits and reservoirs. This is because they have adapted to breed on the shingle banks of rivers that have been scoured clean of any vegetation by winter floods. The shingle,

The ringed plover appears stouter than its smaller relative, and has an orange base to the bill and orange legs.

which is the same colour as the eggs and chicks, camouflages them against aerial hunters such as kestrels.

It was this unusual breeding adaptation that led the little ringed plover to colonise Britain in the years after World War II. With new homes and motorways requiring gravel, quarrying began in several locations around London. The little ringed plover soon found the new habitat to its liking and managed to establish a breeding population, which continues to thrive.

The two species also differ from each other in their migration habits. British breeding ringed plovers go only a short distance south, so Scottish breeders winter on English south coast estuaries. In contrast, little ringed plovers head all the way to sub-Saharan Africa for the winter, usually returning in late March and early April.

Interestingly, both species are now changing their habits. Increasingly, ringed plovers are breeding inland, in the same sort of places as their smaller relative, while in parts of Wales little ringed plovers are nesting on their original habitat of shingle riverbanks. This adaptability is putting them in good stead: both are thriving at a time when less adaptable species are struggling to cope with change.

If you ever come across these birds while they are nesting, you may witness their special 'distraction display'. This occurs when an intruder gets too close to a nest, and the parent bird must take desperate measures to prevent them finding it. It does so by extending one wing, as if it is broken, and hobbling away. At great risk to itself, it lures any predator away from its eggs or chicks.

The golden plover is found on moorland in summer but winters on lowland farmland.

The little ringed plover's yellow eye-ring and olive-green legs distinguish it from its commoner relative the ringed plover.

Lapwing

The lapwing is one of our classic farmland birds, and so has been in trouble for many years now, thanks to changes in farming practices, which have focused on intensive food production to the exclusion of our wildlife. Once flocks of thousands of lapwings were a regular sight almost anywhere in the British countryside; now they have become a rare event indeed.

Likewise, the lapwing was once a common breeding bird, but since 1960 its numbers have fallen by well over four-fifths, and the species is usually only found nesting on specially managed nature reserves. Until we change the way we manage our lowland landscape, and make room for birds such as the lapwing, the decline of this and other iconic farmland species would appear to be inevitable.

This is a pity, because the lapwing is one of our most beautiful waders; indeed, it is among the most striking of all our breeding birds. From a distance they appear black and white, especially when huge flocks wheel around in the sky, alternately flashing their dark upper parts and pale under parts. But get a little closer, and you will see that the lapwing is far from being monochrome: its dark head, back and wings are in fact a complex series of greens, purples and browns, often depending on the angle and intensity of the prevailing light, as the shades are produced by iridescence.

The under parts are mainly snow-white, but with a delicate yellowish-brown patch beneath the tail. But the most obvious feature is the extraordinary crest: a curved spike of feathers sticking up from the back of the crown.

Lapwings look good all year round, and it's hard to beat those impressive winter flocks. But, for many people, the best time of year to watch lapwings is the spring, when males soar up into the sky and tumble back down to earth in an impressive courtship display. All the while they accompany their aerial acrobatics with the most bizarre calls: a series of whoops and squeals memorably described by the television presenter Kate Humble as rather like the soup dragon in the children's TV series *The Clangers*. Another observer has likened them to a tape being played backwards.

Once lapwings have established the pair bond, the real hard work begins. The female lays a neat clutch of four dark, blotched eggs in a shallow scrape among grass in a wet meadow. The eggs are well camouflaged, and they certainly need to be: predators such as crows

and kestrels are ever present, and the parent birds spend much of the time mobbing any intruders to try to keep their precious clutch safe. Like other waders, lapwing chicks are precocial, which means that, almost as soon as they have hatched, they are able to run around and find food. Unfortunately, this places them in great danger, as they are apt to wander away from the nest and their protective parents. Fortunately, when danger looms, they are adept at freezing stock-still, so as not to be seen from above.

Once the breeding season is finally over, lapwings tend to gather on ploughed fields or coastal estuaries, often in large flocks, accompanied by their smaller relative the golden plover. The sound these flocks make is one of the classic soundtracks to any autumn or winter day.

Although this spectacle is becoming increasingly difficult to find in many parts of Britain, there is some hope for the lapwing. Areas such as the Somerset Levels are now being specially managed for breeding and wintering farmland birds, with water levels carefully adjusted and grazing regimes implemented to allow the birds to complete their lifecycle. With luck, the sight of thousands of lapwings on a frosty winter's day, and the sound of their courtship on a wet meadow in spring, will soon become more frequent right across our countryside.

With its iridescent plumage and jaunty crest, the lapwing is one of the most handsome of all British waders.

Small & medium-sized Waders

For many novice birders, waders can appear to be one of the most daunting and difficult groups. Unlike other birds, where colour and pattern are usually helpful in differentiating one species from another, at first sight, waders all appear to be roughly the same shade of muddy brown. The fact that you are usually watching them at some considerable distance, in the dim light of an autumn or winter's day, makes things even more difficult.

Dunlin (above) and Temminck's stint (below) are best told apart by their size and structure: the dunlin is larger with a longer, more decurved bill.

Yet with practice you can get to grips with this group of birds, and soon discover that not only are there clear differences between the species, but that waders as a whole exhibit some of the most fascinating behaviour of all our birds.

The key to identifying waders is first to work out their rough size: small (dunlin, sanderling and stints); medium (sandpipers, snipe and shanks); and large (godwits and curlews – see page 98). Once you have that, you then need to concentrate on the bird's overall shape and especially on its bill shape and size, and the colour of its legs.

The medium-sized redshank, spotted redshank and greenshank are easy to tell apart, mainly because, as their names suggest, they have different leg colours. Redshanks are brash, noisy birds, sometimes called 'the sentinel of the marsh' because they usually sound an alarm call when any intruder approaches. Their legs are orange-red, as is the base of their short, straight bill. But be aware of another wader with similar-coloured legs: the ruff. About the same size as a redshank, it may be mistaken for one, though it has a shorter bill.

The much more elegant greenshanks have, unsurprisingly, green legs and a long, slightly upcurved bill and greyish-green plumage. Both these species breed in Britain: the redshank is quite widespread, whereas the greenshank is mainly confined to the Scottish bog country.

Spotted redshanks are a passage migrant, stopping off on their way to and from their Arctic breeding grounds and their winter quarters in the tropics. They are often seen here in their pale grey winter plumage, with their very long bill and long, blood-red legs making them quite distinctive.

Four species of sandpiper can be regularly seen in Britain: the common, green, wood and curlew sandpipers. Of these, by far the most widespread is the common sandpiper, which is often seen by the edge of any fresh water: a lake, stream or river. Like another riverine bird, the dipper, common sandpipers habitually bob up and down while feeding. This may be something that enables them to confuse their prey or, possibly, to see movement beneath flowing water.

The other three sandpipers are far less common in Britain. All are passage migrants, stopping off here in early autumn as they head south from their north European and Siberian breeding grounds to Africa. Green sandpipers often turn up as early as July, and are usually seen near fresh water. Wood and curlew sandpipers come through in August and September, usually near the coasts.

Finally, the easiest medium-sized wader of all to identify is the snipe. No other British bird has such a long bill for its size, which is used to probe deep into soft mud for its prey.

Identifying the smaller species is much more challenging. A good working knowledge of dunlin, especially in its grey-brown, non-breeding plumage, is essential – before you identify any other small wader, you first need to eliminate this very common species. Look for the shortish, decurved bill – both the large knot and the similar-sized sanderling have straight bills. Sanderlings are easiest to identify by their habits: they run along the tideline like little clockwork toys. If you see a wader that is even smaller than a dunlin, it is most likely to be a little stint, or perhaps the rare Temminck's stint.

The best places to go to see a range of waders are coastal marshes such as Cley and Titchwell in Norfolk or Minsmere in Suffolk, where you will also find plenty of birders who will help you sort out any confusion!

Waders such as the snipe (above left), redshank (above) and common sandpiper (below) have very different shaped bills for different feeding techniques.

Large Waders

The larger the wader, the easier it is to identify, but you may need to ignore their plumage features and concentrate instead on the size and shape of their bills. The curlew and black-tailed and bar-tailed godwits may be encountered in Britain at most times of the year, while the whimbrel, a summer visitor, is not usually seen here from October through to March. All four species are generally found on estuaries and coastal marshes, though whimbrel and black-tailed godwits often stop at inland sites on their journeys north and south to their wintering and breeding grounds.

Only one of the quartet, the curlew, is a common British breeding bird, and it is our largest wader. It once bred on wet meadows throughout the country, but is now mainly confined to moorland habitats in the north, where you can still hear the evocative call that gives the bird its name – this doesn't come, as you might assume, from the shape of its bill.

The curlew's decurved bill is, nevertheless, its most obvious distinguishing feature. Being incredibly long (and even longer in the female), it is used to probe into the deepest mud for molluscs and other items of food. As with many other waders, the tip of the bill is very sensitive, enabling the bird to locate its prey by touch rather than smell or sight.

The whimbrel is basically a smaller, neater version of the curlew, with a slightly shorter bill and distinctive dark stripes across the side of its crown. The best way to tell the two apart is their call: as well as its famous two-note 'cur-lee', the curlew also utters a bubbling sound, whereas the whimbrel gives a distinctive seven-note call.

The whimbrel is a very scarce breeder in Britain, doing so mostly in the far north, with its stronghold in Shetland. However, breeding birds from Scandinavia and Iceland pass through Britain each spring and autumn on their way to and from West Africa, with regular stopping points on coastal estuaries and at inland sites such as the Somerset Levels. Late April is the best time to look for them as they travel through our countryside.

The two godwits – the bar-tailed and black-tailed – are mainly passage migrants, though a few pairs of black-tailed godwits do breed each year in East Anglia and northern Scotland. Superficially, these godwits can appear very similar: both are orange-red in breeding plumage, and brownish-grey in autumn and winter, with long,

The curlew is Britain's largest wader, easily identified by its large size and long, decurved bill.

straightish bills. Since adult and young birds moult at different times, you may encounter a mixed flock with birds in all sorts of plumages.

The best way, though, to tell the two species apart is through their structure: the black-tailed godwit is a tall, elegant bird, with a long neck and a very long, almost straight, bill (with a very slight upturn), while the smaller bar-tailed godwit has a shorter but more upcurved bill and shorter legs. In non-breeding plumage, the black-tailed godwit appears plain greyish-brown, while the bar-tailed godwit's plumage is more scalloped, rather like that of a curlew.

Habitat is another way to tell them apart. Black-tailed godwits usually appear on brackish or freshwater marshes, alongside ducks such as wigeon. The bar-tailed, meanwhile, is a bird of the coast, more likely to be seen feeding along the tideline.

The bar-tailed godwit has one major claim to fame: it carries out the longest non-stop journey of any migratory bird in the world. Birds breeding in Alaska were fitted with special tags enabling scientists to track them on their journey south. To everyone's amazement, they flew non-stop all the way to New Zealand – an epic journey of over 11,000 km (7,000 miles).

The whimbrel is a scarce breeder in Britain, mainly in Shetland.

Black-tailed godwits can be told apart from the bar-tailed by their larger size and longer, straighter bill.

Oystercatcher & Avocet

These two large wading birds have one obvious thing in common – their pied plumage – but their habits and histories as British birds are very different. The oystercatcher is one of our commonest and most widespread waders, and it is also probably the easiest to identify. No other bird has the combination of stocky shape, bold black and white plumage, and, the clincher, that huge, straight, carrot-like bill, used to find food.

Despite its name, the main food of the oystercatcher is mussels, which it finds by probing down into the mud at low tide with its orange bill. Once the waters rise again, as they do twice a day, oystercatchers gather in close-knit, high-tide roosts, sometimes with other wader species such as curlew, knot and dunlin. Here they snatch a few moments of rest, always keeping one eye open for predators such as a passing peregrine, before starting to feed again as the tide drops.

Avocets are also black and white in plumage, but the superficial resemblance between the two species ends there. In contrast to the stocky, short-legged oystercatcher, the avocet is one of our most elegant and graceful birds. Its smart, pied plumage is set off with a pair of long, thin, blue legs and, most obviously, a slender, upturned bill. Like the oystercatcher, the bill is the key to its feeding technique, but rather than probing into mud, the avocet sweeps its bill from side to side as it walks along, picking up tiny aquatic organisms as it does so.

Oystercatchers are noisy birds, often calling from the top of a fence post to defend their nesting territory.

But the avocet is not best known for its habits but for its extraordinary comeback as a British breeding bird. During the Middle Ages, avocets were common in southern and eastern Britain, but the draining of the East Anglian fens and the shooting of the birds for taxidermy meant that the species became extinct here by the middle of the nineteenth century.

For over a hundred years, the avocet was thought to be lost as a British breeding bird until the years just after World War II. In 1947 a couple of pairs returned to nest at two sites on the coast of Suffolk – on farmland that had been flooded to prevent a Nazi invasion. The RSPB wasted no time in buying the land, which they turned into two famous reserves: Havergate Island and Minsmere.

Thus protected, the avocets thrived in their new home, and today they are established as a breeding bird throughout East Anglia. Indeed, their rise in numbers and aggressive habits – avocets will defend their breeding area against any incomers – have led to problems for reserve managers who might want to encourage other birds to nest on their land. In winter, large flocks of avocets can also be seen on estuaries in southwest Britain such as the Tamar and the Exe.

Like the avocet, the oystercatcher is mainly found at coastal sites, along shingle beaches and on marshes. But both species do occasionally venture inland to breed on gravel workings or nature reserves: the avocet has even, on occasion, nested in the London area. Both avocets and oystercatchers appear to be bucking the trend for many of our breeding birds and are on the increase. Oystercatchers can be seen around most of our coasts, especially on estuaries and mudflats. In Scotland they are a common sight along the side of roads – not just by the coast but also inland, by the edges of lochs, where they also breed.

The story of the avocet's return as a British breeding bird became so well known that soon afterwards the RSPB adopted the bird as its symbol, which it remains to this day. And it is certainly true that if you want to get close-up views of avocets, the best places to do so are still the RSPB reserves of Havergate Island and Minsmere in Suffolk, and Titchwell on the north Norfolk coast.

With its black and white plumage and upcurved bill, the avocet (above and below) is very easy to identify.

Rails

The rails are an amazing family. They are found all over the world, even on offshore islands that very few other birds manage to reach, yet when they fly, they look clumsy and awkward, as if their short wings could hardly carry them more than a few metres. It is almost impossible to believe that they travel such vast distances.

Of the four main British species of rail, only one, the corncrake, migrates away from our shores. After breeding, this elusive bird flies south, across Europe, the Mediterranean Sea and the Sahara desert, to spend the winter somewhere in Africa. When they return here in spring, they are almost impossible to see, as they skulk away to hide in deep vegetation in overgrown fields in their breeding strongholds of the west coast and islands of Scotland, and parts of rural Ireland.

The water rail (above) and corncrake (below) are both shy and elusive, rarely giving good views.

After decades of decline caused by the replacement of traditional crofting with modern farming methods, the corncrake is finally making a comeback. Farmers have been encouraged to manage their land in a corncrake-friendly way, and today their monotonous, repetitive call can be heard throughout the Western Isles. There is even a small, reintroduced population of corncrakes in East Anglia, giving hope that one day, perhaps, the species will be found throughout rural England and Wales, as it was only two centuries ago.

The corncrake is sometimes known as the 'land rail', to distinguish it from its more aquatic cousin the water rail. This bird gave its name to the saying 'thin as a rail' – its body is so slender that it can squeeze between reeds in its watery habitat. Water rails are just as elusive and hard to see as corncrakes, and, like them, they are best located by their call – a series of high-pitched squeals sounding rather like a piglet in distress.

Water rails are especially vulnerable in hard winter weather, as ice means they are unable to find any food. This does, however, often force them to become bolder, which allows us to glimpse this attractive and striking bird. Look out for a long-legged, pot-bellied bird with a long, red, dagger-like bill and stripy pattern down its flanks.

The corncrake and water rail are elusive and difficult to see, but the same cannot be said of the other two members of the rail family that commonly breed in Britain: the moorhen and the coot. Unlike other rails, these two species swim like ducks and, indeed, are often found alongside them on park ponds, lakes and rivers.

People often get confused when trying to tell moorhens and coots apart, but it is, in fact, very simple. Coots are all black (actually very dark grey), apart from a prominent white bill and shield on the front of their face, giving rise to the saying 'as bald as a coot'. Moorhens are smaller and more slender, and have a purplish-brown plumage, with a thin, ragged, pale line along their flanks and a bright red and yellow bill. The only real confusion comes with baby coots, which, strangely, have a red bill, rather like that of a moorhen.

The moorhen is one of our most familiar waterbirds – its name derives from the word 'mere', meaning small lake or pond.

Their habits are different, too. Moorhens are often found on tiny patches of water which other birds would ignore, such as narrow ditches, small ponds and dykes. They hardly ever fly, though they are known to take to the wing at night to move between one area and another. Coots prefer larger expanses of water, especially in winter, when thousands of them may gather at a reservoir or lake, to dive for food. Moorhens, by contrast, either pick items from the surface of the water or find food on land, usually on areas of damp grass near water. When they do so, you can see that their feet are not webbed, like ducks, but instead have separate toes.

Both moorhens and coots build a floating nest of aquatic vegetation, usually quite close to the shore, which may allow you to watch as they incubate their clutch of eggs. Once the chicks have hatched, they take immediately to the water and find their own tiny morsels of food straightaway.

Coots rarely take to the air, usually only flying a short distance just above the surface of the water before landing again.

Dabbling Ducks

Our various kinds of duck fall into three rough categories, depending on their feeding technique and habitat. These are 'dabbling ducks', 'diving ducks' and 'sea ducks', though almost any species, including the familiar mallard, can sometimes be seen out at sea.

There are seven species of dabbling ducks found in Britain: the mallard, teal, shoveler, wigeon, gadwall, pintail and garganey. They have been given the name 'dabbling' because they mainly find food when sitting on the surface of the water, either dipping their bills beneath or 'upending', rather than diving. By far the commonest and most familiar of them is the mallard. The handsome males, with their bottle-green head, and drabber females, are a familiar sight to anyone who has ever fed the ducks at their local park pond, lake or river.

Mallards are a common breeding bird in Britain, and also found in large numbers in winter. However, they are often most obvious in spring, when the males can get very noisy and aggressive, pursuing an unfortunate female in gangs of three or four, and even forcing her underwater in their desperate attempts to mate.

While the mallard is one of our largest and showiest ducks, the teal is one of the smallest and hardest to see. Shy and flighty, teals are often found in marshy areas of water such as dykes and ponds, where they hide away in the vegetation. You may approach them very closely before you realise they are there, and often the first indication of their presence is when a tight-knit group shoots up into the air on their pointed wings before flying rapidly away. No wonder the collective name for this bird is a 'spring' of teal. The male is one of our most beautiful ducks, with a combination of pearl-grey, chestnut and green plumage. As with other ducks, the female is much less brightly coloured, as she needs to be safe from predators such as foxes when sitting on her nest.

Male gadwall (above) and teal (below) are much more brightly coloured than the browner females, which must stay camouflaged as they sit on their nest.

The shoveler, wigeon, gadwall and pintail are found in Britain all year round, though they are far more numerous during the autumn and winter months, when their numbers are augmented by birds from the north and east of the British Isles. For example, only a couple

of hundred pairs of wigeon actually breed in Britain, mainly in Scotland, whereas more than half a million birds come here in autumn from Siberia, to take advantage of the mild winter climate and abundant supplies of food.

Seen well, the males of these four species are easy to identify, each having obvious field marks. Look out for the huge bill of the shoveler, and its white breast, green head and chestnut sides; the pearl-grey plumage of the gadwall; the yellow face and chestnut head of the wigeon; and the long tail of the pintail. Females are trickier, but shape becomes important here: each species has a subtly different 'jizz', as birders call their overall appearance. But be aware that in late summer, the moulting season, the males of any ducks go into what is called their 'eclipse' plumage, and look very similar to their drabber mates. Only when autumn arrives do they transform back into their handsome garb.

These four species also have rather different feeding habits. The gadwall and pintail feed fairly conventionally, sitting on the water and picking off items of food, or upending to take food from just beneath the surface. Interestingly, gadwalls often associate with coots, perhaps because the coots' diving brings food up from the bottom of the lake. Shovelers often feed communally, with a large group gathering closely together, using their huge bills to grab whatever they can find. Wigeons, on the other hand, have small, neat bills and use them not on the water but to graze short grass, neatly picking off the tips as they walk deliberately across a wet meadow.

The seventh dabbling duck, the garganey, is much scarcer than its relatives, and it is only a summer visitor, spending the winter in Africa and returning to Britain in March and April. Even then they are a very rare breeder and are seen much less often than the other dabbling ducks. This is a pity, as the male garganey is arguably our most handsome duck, with a beautifully marked plumage of browns, creams and greys.

Wigeon (left) and shoveler (right) travel in large flocks from Siberia to spend the winter in Britain, whereas garganey (below) migrate to Africa for the winter.

Diving Ducks

Diving ducks, as their name suggests, have a very different feeding technique from dabbling ducks. Instead of upending gracefully or nibbling at the surface, they plunge right down beneath the water to get their food, using their powerful webbed feet to propel themselves. Diving ducks can be found in a wide range of habitats, from gravel pits and reservoirs to lakes and the open sea, though they tend to avoid small areas of shallow water, where diving would be difficult.

The two commonest freshwater species, the tufted duck and pochard, both breed in Britain, but, like the dabbling ducks, their numbers are hugely increased each autumn as hundreds of thousands of birds head south and west towards the British Isles from Scandinavia and Siberia, to enjoy the benefits of our ice-free waters.

Both the tufted duck and pochard are fairly easy to identify – or at least the males are. The male tufted duck is basically black and white: black on the head, breast and back; white on either side; with a small tuft of black feathers, which gives the species its name, protruding from the back of its crown. Male pochards are equally striking, with their deep chestnut head, grey back and wings, and black breast. Females are much trickier to tell apart: female tufted ducks are basically dark brown, with lighter flanks and a small tuft on the head, while female pochards are paler and a greyer brown, with a characteristic long forehead and bill. Both tend to gather in large flocks, where males and females are together, which makes identification easier.

Pochard (above left), smew (below) and goldeneye (above right) are all winter visitors to Britain; pochard and goldeneye also breed in small numbers.

Three other diving ducks – the goldeneye, goosander and smew – may also be seen in winter on our lakes, gravel pits and reservoirs. They are very handsome and striking birds, and, provided you have good views, are easy to tell apart, especially the males. The goldeneye male is mainly black above and white below, while the male goosander has a dark green head and pinkish-white body. The females of both species are grey with chestnut heads. These two birds breed in northern Britain as well; goldeneyes, in particular, have taken to nest boxes since they first colonised Scotland from Scandinavia in 1970, and are now thriving as a result. In early spring, flocks of them often congregate on reservoirs, where the males display to the females, throwing their head backwards in a

spectacular attempt to win a mate before they head back north to breed.

Smews do not breed in Britain, and are much scarcer than their larger relatives. They are normally only seen at well-known sites in southeast England such as Dungeness in Kent or the gravel pits around west London, though if there is hard weather on the near continent, they may arrive in greater numbers than usual. The male smew is a stunning combination of white with grey and black marks; the female is grey with a chestnut crown and white cheeks. Both are small, about the size of a teal.

Other kinds of diving duck are mainly found offshore, as they prefer to spend the winter months at sea. Of these, the commonest is the eider duck, which also breeds around our northern coasts. Male eiders are very handsome birds: stocky and black and white, with a pinkish flush on the breast, and green on the side of the head. When breeding, they also make an extraordinary sound, rather like the late comedian Frankie Howerd! As with other ducks, the females are much less colourful – a subtle mixture of browns enables them to be camouflaged on their nest. Once their ducklings have hatched, eiders often gather together in 'crèches', with several females looking after one another's offspring.

Eider (above) and goosander (below) are often found off our coasts or in harbours and bays.

The best time to look for any of these diving ducks is during the autumn and winter months, and usually on large areas of open water such as reservoirs, estuaries or the open sea. They will often be at quite a distance from you, so you may need a telescope. During the breeding season they can also be seen along rivers in upland areas of northern Britain (especially the goosander and red-breasted merganser), or on rocky shores and lochs (the eider and common scoter).

Geese

Wild geese have a magic all their own, and have inspired generations of artists, writers and naturalists, including the patron saint of the British conservation movement, Sir Peter Scott. This is largely due to their habit of gathering in huge flocks, often in some of Britain's most beautiful wild places, such as windswept estuaries and coastal marshes, in what remains one of the greatest British wildlife spectacles. How odd, then, that most people's encounters with geese tend to come while feeding the ducks in their local park, when a boisterous Canada goose will muscle in and steal the food intended for their smaller relatives.

Feral populations of Canada (right), Egyptian (below) and greylag (bottom) geese can all be found in southern Britain.

As their name suggests, Canada geese don't really belong in Britain. They were introduced here from their native North America several centuries ago and, in the absence of competitors or predators, have thrived to the point at which they are now considered to be a pest. Attempts have been made to cull them, as their droppings can cause damage to grass in parks, and they even eat the offspring of other waterbirds such as mallards, but so far their rise seems set to continue. In the past couple of decades, Canada geese have been joined by two other feral goose species: Egyptian geese, originally from Africa, and greylag geese, which do occur naturally in northern Britain but have now established feral populations in the south.

It is fortunate that we also have several species of truly wild geese. The native greylag, along with five winter visitors to our shores – brent, barnacle, white-fronted, pink-footed and bean geese – all come here in autumn to feed in our fields, salt marshes and estuaries.

Wild geese have had their ups and downs, not least because farmers often regard them as a pest. Having thousands of geese land on your field of winter wheat or sugar beet is no joke: as well as eating the crop, they also trample the fields, causing major problems for farmers. Fortunately, many are now compensated for this, and geese and humans are beginning to live in closer harmony than before.

As a result, wild geese are doing rather well nowadays. Pink-footed geese, in particular, have seen a very rapid rise in numbers, which is crucial, as almost the entire world population of this attractive goose spends the winter months in Britain. Flocks of tens of thousands of them can now be seen in their north Norfolk stronghold, especially at dawn and dusk, when they fly from their roosting areas to the fields to feed.

Brent geese are also doing very well. This small, dark goose comes here from both the northeast and northwest, with some travelling from as far afield as the Canadian Arctic. They are often seen on our south coast estuaries, where they feed on eelgrass, or in fields alongside roads. Once thought to be endangered, brent geese have increased hugely in numbers in the past few decades.

White-fronted geese also travel a long way to get to Britain. One population comes from northern Siberia, and winters mainly on coastal marshes in southern and eastern England, while another travels all the way from Greenland to the Scottish island of Islay, home of malt whisky. There they feed alongside huge flocks of barnacle geese from Spitsbergen (Norway). Wild greylag geese can also be seen in Scotland, their honking calls often giving away their presence before you see them.

The rarest British goose, the bean goose, is far harder to see. Indeed, your best bet is to visit the regular wintering flock in the Yare Valley in east Norfolk. Other good places to see wild geese include the WWT (Wildfowl & Wetlands Trust) centres at Slimbridge in Gloucestershire, Caerlaverock on the Scottish borders, and Castle Espie in Northern Ireland.

Pink-footed (above) and brent (below) geese come here from the north and east in autumn to take advantage of our mild winter climate.

Swans

The sight of wild swans – whooper and Bewick's – on a chilly January day is one of the iconic spectacles of the British winter, while the regular gathering of mute swans, our resident species, on a summer's day along the banks of a river is just as memorable.

The mute swan holds a special place in our heritage, and is also the one we are most familiar with. They are our largest breeding bird and one of the first we get to know as children, as a stately white creature looking on while we feed bread to squabbling ducks. Yet, despite our familiarity with mute swans, we have managed to create all sorts of myths around them.

It is often said that all swans belong to the Queen but, in fact, those on the River Thames may have different owners, and there are many swans elsewhere in the country that do not belong to any individual. Another famous myth is that 'a swan can break a man's arm'. While it is true that male swans, in particular, can get very aggressive, especially if you approach their nest or cygnets during the breeding season, they certainly don't have the strength to cause such serious injury. The belief that all swans pair for life is generally correct, though, of course, if an individual dies, its partner will seek out a new mate. Finally, the name 'mute' swan belies the fact that they make a range of noises, including a loud hissing!

Mute swans are very easy to identify, thanks to their large size, all-white plumage and orange bill with a black knob above – this is larger in the male (known as the cob) than the female (the pen). Cygnets are, of course, the original 'ugly duckling' of the Hans Christian Andersen fairy tale. When they hatch, they are covered in soft grey down, which is followed by a blotchy grey and white plumage until they finally moult into their full adult garb a year or so after they hatch.

The two 'wild swans', whooper and Bewick's, are superficially quite similar: both have a yellow and black bill. Bewick's swan is considerably smaller, though this can be hard to judge on a lone bird without the other species alongside for comparison. The best way

Bewick's swans (above) and whooper swans (opposite) can be told apart from each other by the whooper's larger size and greater extent of yellow on its bill.

to identify them is to look more closely at the shape and colour of the bill: whooper swans have a long bill, which appears more yellow than black; Bewick's, meanwhile, have a shorter, more compact bill, which is more black than yellow. Over half a century ago, the great conservationist Sir Peter Scott realised that individual Bewick's swans could be identified by the unique pattern on their bill. This enabled him to begin a close study of those birds wintering at the WWT (Wildfowl & Wetlands Trust) HQ at Slimbridge, a study that continues to this day. Incidentally, their name commemorates one of the great pioneers of British ornithology, the engraver and publisher Thomas Bewick, whose *History of British Birds*, published at the turn of the nineteenth century, was one of the first popular bird books.

Whooper and Bewick's swans both arrive in Britain in mid-autumn, the whoopers mainly coming from Iceland, and the Bewick's travelling a rather longer distance, from Siberia. In some years, fewer swans come here – if the weather conditions remain mild in continental Europe, they may remain there throughout the winter. They usually stay in Britain until February or March, when they pair up, often performing elaborate and noisy courtship displays, before heading back north and east to breed.

The best places to see wild swans are at the WWT centres at Slimbridge (Bewick's), Caerlaverock (whoopers) and the Ouse Washes in Cambridgeshire (both species). Stay until dusk and you will see them being fed and have really good close-up views of these elegant birds.

Mute swan cygnets regularly
hitch a ride on their parents' backs.

Pelagic Seabirds

Leach's petrel (above) and storm petrel (opposite) are two of the most mysterious and enigmatic of all Britain's seabirds, rarely seen close to shore except after stormy weather.

One of the greatest challenges facing a birder in Britain is catching sight of the most elusive and mysterious group of all our birds, the so-called 'pelagic' seabirds: shearwaters and petrels. Pelagic derives from the Greek word *pelagos*, meaning ocean, and these birds truly are birds of the high seas. They spend the vast majority of their lives riding the waves, coming to land only when they must, in order to breed.

Three species of shearwater and petrel regularly breed in Britain: Manx shearwaters and storm and Leach's petrels. Of these, by far the most numerous is the Manx shearwater, a long-winged, black and white seabird that, as its name suggests, flies low over the surface of the waves on stiff wings. Their name comes from one of their breeding colonies on the Calf of Man (the southern tip of the Isle of Man), but their main breeding areas are the Scottish island of Rum, and the Welsh islands of Skomer and Skokholm, off the Pembrokeshire coast. Here they nest in old rabbit burrows, laying a single egg and spending several months feeding their chick until it is large enough to leave the nest and fend for itself.

Manx shearwaters will only return to their nests after dark, ideally on a cloudy night. If they attempt to do so before dusk falls, or on a moonlit night, a predatory great black-backed or herring gull may pick them off. To have a chance of seeing these amazing birds, your best bet is to take a boat trip around the islands on a fine evening in midsummer – from May through to August – when they will often gather in vast numbers offshore, allowing you to get close views of them. If you are lucky enough to visit a shearwater colony after dark, you will hear the most extraordinary cacophony of blood-curdling sounds, rather like the wailing of a particularly demented banshee!

The two other members of this group – storm and Leach's petrels – are our smallest breeding seabirds. Storm petrels are barely larger than a house martin and, with their black plumage and white rump, they do actually resemble them a little. Although storm and Leach's petrels look very similar superficially, they can be told apart both by their size difference – Leach's is noticeably larger – and by their flight action. Storm petrels flutter rather like an oversized moth, while Leach's have

a much stronger flight action, more like a tern or even a nightjar than their smaller relative.

Storm and Leach's petrels breed in colonies, mostly off the north and west coasts of Scotland, on islands such as North Rona, the Flannan Isles and St Kilda. Like Manx shearwaters, they only venture onto land after dark, each uttering their bizarre calls – the storm petrels are said to sound like 'fairies being sick', while Leach's petrels sound like an amusement arcade game. The call of storm petrels also echoes around one of our oldest buildings, the Iron Age broch (round tower) on the island of Mousa in Shetland, where the birds nest in tiny cracks in the ancient walls.

Seeing these birds is very difficult. Probably the best way is to take a ferry across either the Irish Sea or the Minch (between western Scotland and the Outer Hebrides), or to go on a specially organised 'pelagic trip' off West Wales or Cornwall. The only time petrels come close inshore is during heavy autumn gales, when you may catch sight of one battling the winds from a west coast headland, or, if the gales are particularly strong, flying around an inland reservoir.

Autumn is also the best time to catch up with even rarer seabirds, including great and sooty shearwaters, which travel all the way from their breeding grounds in the South Atlantic Ocean, and occasionally pass by headlands such as Flamborough Head in Yorkshire or St Ives in Cornwall.

Gannet, Cormorant & Shag

Of all Britain's two dozen or so species of seabird, without doubt the largest and most impressive is the gannet. Watching gannets plunge-diving into the sea, folding back their wings in order to enter the water at high speed to catch fish, is one of the greatest of all our wildlife spectacles. And seeing them in their vast colonies, numbering tens of thousands of individual birds, is a once-in-a-lifetime experience.

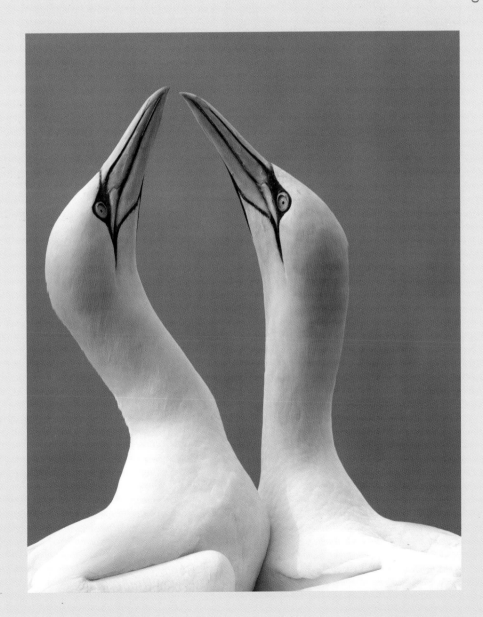

Gannets have a very strong pair-bond, which they maintain through ornate courtship rituals.

Cormorants now nest inland far more frequently than they used to, usually building their nest in a dead tree.

Britain supports almost three-quarters of the entire world population of gannets, the remaining birds being on the other side of the North Atlantic. The largest British colonies are on the island of Boreray (part of the St Kilda group), Grassholm off the Pembrokeshire coast, and, most famously, Bass Rock in the Firth of Forth near Edinburgh. As you approach, this vast rock appears to rise from the sea, surrounded by thousands of gannets in the air all at once, making a cacophony of sound. Only when you get closer do you also notice the smell of guano generated by so many birds.

Gannets can also be seen off any of our coasts, at any time of year. From a distance, they may at first appear like large gulls, but no other seabird has the gannet's unique combination of long, pointed, white wings tipped with black, a dagger-like bill and a yellow patch on the crown. Young gannets may be harder to identify, as they show varying amounts of black on the wings. Nevertheless, they always appear bigger than even the largest gull.

Unlike many of our seabirds, gannets do not appear to have suffered from the food shortages that have led to population declines for other species such as the puffin and guillemot. This may be because gannets feed on larger kinds of fish, which are still plentiful in the seas around our coasts.

Two other seabirds, the cormorant and shag, are frequently seen alongside gannets. Both are much smaller and mainly dark in colour. The cormorant is the larger of the two, with a distinctive yellowish patch around the face, and it often also shows a large, pale patch on its under parts. Shags are a dark greenish-black, with a bright yellow bill and, during the breeding season, a distinctive crest that gives the bird its name.

The other main difference between cormorants and shags is their preferred habitat. Shags are almost exclusively found on the coast and are commoner the farther north you go, whereas cormorants, though once a mainly coastal bird, have in the past few decades headed inland. In recent years, birds from the near continent – a different race from our native birds – have also begun to breed in colonies in the south and east of England. They can be told apart by the greater amount of white in their plumage.

Today, cormorant colonies are just as likely to be in trees around your local gravel pit as by the sea; this is the consequence of the shortage of fish in our seas combined with the greater availability of food inland. This move inland has meant that cormorants have come into conflict with anglers, who are concerned that these birds are taking 'their' fish. This echoes similar conflicts in the past around our coasts, where bounties were often put on the heads of these birds in order to keep numbers down.

Unlike other seabirds, which have special oil glands enabling them to waterproof their feathers for a sea-going existence, cormorants need to dry their feathers regularly to avoid them getting waterlogged. This explains why you often see cormorants holding their wings outstretched, a position that also helps them digest their food more rapidly.

The shag's crest and lemon-yellow patch at the base of its bill are good ways to distinguish it from its larger relative the cormorant.

Herons & Egrets

Until relatively recently, only two species of long-legged waterbird bred in Britain: the common and widespread grey heron and the rare and elusive bittern. How things have changed. During the past decade or so, these have been joined by half a dozen other species, together with several others on the verge of breeding here.

The most obvious of these new arrivals is the little egret. This elegant, Persil-white heron was once regarded as a bird of southern Europe, which you would only see on holidays to the Mediterranean region. Then, from the late 1980s onwards, birds from western France started to head north after breeding, and flocks of little egrets – once a major rarity – began to be seen regularly on south coast estuaries and nature reserves. The invasion had begun, and within a few years little egrets had not only started to breed, but had spread north to many parts of England and Wales. Today, almost one thousand pairs of this attractive bird breed every year, usually in heronries alongside their much larger relative the grey heron. Little egrets can now be seen on marshes, estuaries and other wetlands throughout southern Britain, even occasionally in the centre of London along the River Thames.

Recently, two other close relatives have joined the little egret: the smaller cattle egret and the larger great white egret. Cattle egrets are most familiar from wildlife documentaries set in the African savannah, where they ride the backs of elephants, so what are they doing here? They are simply continuing the huge global expansion of one of the most successful of all the world's bird species. Having colonised much of the New and Old Worlds from their original home in Africa and Asia, they are heading north into Britain. Cattle egrets have now bred here on several occasions and look set to become a permanent colonist in the very near future. Their much larger relative, the great white egret, may soon join them. This bird has also expanded its range northwards in Europe, and is now a regular sight in wetlands such as the Somerset Levels.

Several species of heron now breed in Britain, including the longstanding grey heron (above) and new arrivals the little bittern (below right) and little egret (bottom).

Other exotic European waterbirds that may also colonise Britain are the little bittern, spoonbill and purple heron, all of which have bred here in the past decade. These species are being helped by two factors: global climate change, which enables them to expand their ranges towards the north; and habitat restoration, especially of wetlands, which gives them a place to breed when they get here.

Habitat restoration has also been the key to the success of the crane, another long-legged waterbird unrelated to the heron but similar in appearance. Cranes have not only been reintroduced in the West Country, but have also managed to colonise under their own steam, with breeding birds in several parts of East Anglia. Seen well, they are unmistakable: tall, stately, grey birds, with black and red heads and a flurry of feathers at the rear.

Our two original species, the grey heron and bittern, are also thriving. Grey herons are one of our earliest breeders, building their huge stick nests in noisy colonies from January onwards. London's Regent's Park is one of the best places to watch their antics and get close-up views of some fascinating breeding behaviour. Bitterns, meanwhile, have expanded their range westwards and northwards from their East Anglian stronghold. The best time to look for them is either in early spring, when the males utter their extraordinary booming call, or midsummer, when they are feeding their young and may be seen flying to and from their nest, hidden inside the reed bed. Their folk-name, 'butter bump', derives from their deep, booming call, which can be heard up to 5 km (3 miles) or so away.

Bitterns do, however, have an Achilles heel: they are very vulnerable to spells of freezing winter weather, when iced-up waters make it impossible for them to find their aquatic food. This misfortune may, however, give you the chance of seeing them, as they often forsake their reedy hiding places out of sheer desperation and feed out in the open.

Once incredibly rare in Britain, sightings of great white (above) and cattle (below) egrets are on the rise.

Kingfisher & Dipper

Some birds are generalists, able to live their lives in a range of habitats; others have chosen to specialise, finding a specific niche, which they then exploit to the full. Of the latter group, perhaps the finest exponents are two waterbirds: the kingfisher and the dipper. Both have evolved a range of behaviours and techniques to survive and thrive in one of the most specialised habitats of all: streams and rivers. In the case of the dipper, this is even more extraordinary, given that it is a songbird, related to wrens and thrushes. It is the only British songbird to live such an aquatic lifestyle.

The kingfisher is seldom seen, but, once glimpsed, cannot be ignored. It is by far the most colourful and striking of all our native breeding birds – only its exotic continental relatives the bee-eater and roller rival it in terms of dazzling plumage.

The kingfisher is often described as a 'blue' bird, but to do so is to ignore two things: first, the extraordinary range of different shades of blue it manages to display, from deep electric-blue through azure to a virtual green; and second, the deep orange hue of its underparts. The other thing that surprises people when they see a kingfisher for the first time is its size: they are tiny, not much larger than a sparrow.

Surprisingly for such a colourful bird, the first indication of a kingfisher's presence is often their sound: a high-pitched series of calls that echoes along a waterway as the bird flies away from you. Getting close to a kingfisher requires a lot of patience; often sitting still and quiet after you hear it call is the best way to see it, as they tend to patrol up and down along the same length of river. Also, look out for nest holes in spring. These are generally in sandy banks, where the birds can make a burrow to keep their eggs and chicks safe from predators.

The highlight of any encounter with a kingfisher is seeing it do what it does best: catching fish. They plunge down from a perch into the water to grab their unsuspecting prey, then return to a perch, where they bash the fish until it is stunned or killed. Sometimes they will not eat their catch but present it to their mate as a love token. Incidentally, you can tell male from female kingfishers by the colour at the base of the bill: orange in the female, black in the male.

The dipper is Britain's only aquatic species of songbird, specially adapted to life in rivers and streams.

The dipper is far less colourful but just as fascinating to watch. Unlike other riverine birds such as the grey wagtail (see page 34) or common sandpiper (see page 96), it does not simply feed by the water's edge, but plunges beneath the surface and either swims or walks along the bottom in search of fish or aquatic insects. Underneath the water, dippers lose their chunky appearance and become slender and graceful. Above water, they usually perch on a rock and bob up and down. No one is quite sure why they do this but it must help them either avoid being seen by their prey or perhaps get a better view beneath the water.

Seen well, dippers are unmistakable. No other bird has the combination of stout shape, cocked tail and dark plumage with that bright white bib. Youngsters leave the nest before they are fully ready to live on their own, and may still be fed by the adults; look out for a yellowish and brown bird perched by the side of the water, often calling furiously to remind its parents that it is hungry.

Dippers are mostly birds of fast-flowing streams and small rivers, where the water is oxygenated enough to be home to plenty of aquatic invertebrates. The best places to look for them tend to be in the north and west, along hill and mountain streams. Kingfishers are more widespread, and in cold winter weather may even be seen on coastal estuaries, where the water remains unfrozen.

Few, if any, British birds can rival the kingfisher for its colour and beauty.

Grebes & Divers

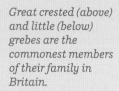

We often think of seabirds as our most water-loving group of birds, living as they do for the majority of the year out on the open ocean. But two other families of birds – the divers and the grebes – not only live their entire lives on the water, but they also make their floating nest there. This gives them a good claim to be the most aquatic of all our birds.

Of the eight regularly occurring British species (three divers and five grebes), by far the best known is the great crested grebe. This is partly because of its beauty – in breeding plumage, it boasts elegant tufts of feathering, which complement its slender body – but also because of its extraordinary history. The great crested grebe was at the centre of the controversy over the welfare of our birds that eventually led, towards the end of the nineteenth century, to the founding of the RSPB. At the time, grebe feathers were used as adornments in women's fashions and, as a result, the species had become almost extinct here. Fortunately, thanks to the efforts of a determined group of (mostly women) campaigners, the grebe was saved.

Great crested (above) and little (below) grebes are the commonest members of their family in Britain.

Great crested grebes enjoyed a boom in the years following World War II, thanks to the creation of gravel pits – artificial lakes, dug to obtain sand and gravel for building roads and homes. They are also frequently found on rivers and lakes, mainly in southern Britain.

The great crested grebe is the largest British representative of its family, with the little grebe, or dabchick, the smallest, being barely bigger than a mallard duckling. As its alternative name suggests, it looks rather like a baby waterbird, with a powder-puff of feathers at the rear, and a tiny bill, which gives it a rather endearing appearance. In breeding plumage, it adopts a splendid chestnut garb, with a tiny luminous green spot near the bill. Little grebes, rather like moorhens, prefer small areas of water such as ponds and ditches, where few other waterbirds would attempt to live. Their presence is often given away by their bubbling call.

Three other kinds of grebe are winter visitors and breed rarely in Britain. The black-necked and Slavonian grebes, which are rather similar in both breeding and non-breeding plumages, are best told apart by their shape and structure. The black-necked is smaller and plumper than the Slavonian, with a fluffy rear end; the Slavonian is far more sleek and more slender. Both have orange tufts behind their eyes in summer and are basically grey, black and white in winter.

Black-necked grebes breed in a few sites around the country, but are easiest to see in autumn and winter, when they head for reservoirs or the coast. Slavonian grebes breed only in the Scottish Highlands, but are regular visitors around most coasts and harbours outside the breeding season. The breeding plumage of the red-necked grebe is rarely seen in Britain, which is a pity as it is rather splendid. In winter, they look more like young great crested grebes, and can be hard to identify.

All the grebes build a floating nest out of vegetation, and have two unusual habits. They cover their eggs with waterweed in order to hide them from predators, which turns them a greenish shade, and once the young have hatched, they often hitch a lift on the parents' backs, hiding away in their feathers so you only realise they are there when they jump off!

The three divers are also birds of one habitat in summer and another in winter. The two smaller species, the red-throated and black-throated, are confined to Scotland as breeding birds. The stronghold of the red-throated is Shetland, where it nests by the side of lochans, only just managing to emerge from the water before flopping down on its clutch. There it is known as the 'rain goose' because its haunting cry is supposed to foretell the coming of bad weather. Black-throated divers are far harder to see, as they breed on remote highland lochs, while great northern divers do not breed at all in Britain.

Divers are often easier to see in autumn and winter, either just offshore or, on occasion, feeding on reservoirs. In breeding plumage, they are very distinctive, but outside the breeding season they can be very hard to tell apart. The posture, and the shape of the bill are often the best means of identifying them.

Red-throated (above) and black-throated (left) divers both breed in parts of Scotland, usually on the edge of freshwater lochs.

The Slavonian grebe (below) is a rare breeding bird, mainly in lochs in the Scottish highlands.

Mammals

Fox

Love them or hate them, it is hard not to agree that the fox is one of our most handsome mammals.

Few British mammals divide opinion quite so much as the fox. Many people admire foxes for their handsome appearance and cunning habits; others loathe them, often for that very cunning that enables them to raid henhouses and dustbins for food.

Our attitudes to foxes also vary depending on whether we live in the town or the country. In some ways this is because urban and rural foxes behave almost as if they were two different species.

Town and city foxes are brash and confident, and appear to have no fear of humans, staring you down if you happen to come across them as they go about their daily business. Country foxes, on the other hand, are shy, wary creatures; if they catch a glimpse of you, they will usually turn and run away. They have good reason to do so. Despite the recent ban on hunting with hounds, foxes are still persecuted by farmers, gamekeepers and landowners, who regard the creature as their sworn enemy because of its highly efficient killing skills.

As one of our largest terrestrial carnivores – only badgers and otters are heavier – foxes thrive across most of lowland Britain, a testament to their adaptability. They can live in woodland, farmland and scrub – almost anywhere they can find food and a place to shelter and raise a family. Some foxes have used their ability to take advantage of our wasteful habits to carve out a unique niche in the leafy suburbs of cities such as London and Bristol, where there are still plenty of woodland and parkland habitats.

The key to the success of foxes is their wide-ranging diet. They are opportunistic feeders, whose prey includes rabbits, small mammals such as mice and voles, invertebrates (including beetles and earthworms) and windfall fruit, as well as domestic animals such as ducks and hens. Foxes are also scavengers – of any dead creatures they come across, and also of waste food taken from rubbish bins outside offices, restaurants and our homes.

Another reason foxes are so successful is that, apart from humans, they have very few predators. But in certain areas, especially game-shooting parts of East Anglia, persecution by farmers and gamekeepers has been so effective that they are now very scarce.

If we can overcome our prejudice against the fox, however, it is a fascinating creature, especially in its breeding behaviour. The breeding cycle begins late in the year, when males defend their territories by calling loudly to one another after dark. Females are attracted by the calling too, and mating generally happens between December and February. The female will seek out a safe place to give birth, either in a specially dug burrow or in an enlarged rabbit hole or old badger sett. In towns, foxes often make their den underneath a garden shed, or in a quiet corner of a churchyard or cemetery.

The litter of four or five cubs is born in spring, usually between March and May, and from May onwards you may catch sight of them as they emerge from the den for their first experience of daylight. They can be very active at this stage of their lives, indulging in play-fighting and acrobatics with their siblings. But they soon gain their independence, being weaned after about a month, and leaving the area where they were born by the autumn.

Culturally, foxes have a long association with humans, having appeared in early writings such as Aesop's *Fables*, and featuring prominently in children's stories. These usually celebrate the fox's cunning, though they also portray the creature as a ruthless killer – testament to our continued conflict over one of our most fascinating yet controversial mammals.

Foxes are among our most adaptable mammals, enabling them to colonise a wide range of urban and rural habitats.

Badger

After the extinction of the wolf, lynx and bear, the badger took on the role of Britain's largest terrestrial carnivore. At around 10–12 kg (22–27 lb) in weight, and almost one metre (yard) in length, the badger is a formidable creature and the most instantly recognisable of all our mammals. No other creature has the combination of black and white head pattern, rotund shape and shuffling gait – despite being related to otters and stoats, the badger has none of their grace and sleekness.

Badgers are creatures of the countryside. Although they are sometimes seen in our towns and cities, these urban populations are a relic of the past, rather than new colonists. Today they can be found in a range of wooded and farmland habitats throughout Britain but excluding most of our offshore islands, upland regions and the far north of Scotland. They are especially common in the south and west of England.

These highly social animals live together in groups of up to 25 adults, with their cubs, in a series of tunnels, known as setts. Some of these may be several hundred years old, but they have, of course, undergone many changes over time, with new holes and tunnels being dug as and when they are required. Oddly, the badger's reputation as a social creature is a particularly British phenomenon – those in continental Europe are far more solitary. One theory suggests this is down to our wetter climate, which allows British badgers to find and dig for food successfully within a smaller area than their European counterparts, enabling them to live in denser and more sociable concentrations.

Badgers are, like so many mammals, mainly nocturnal in their habits. They usually emerge from their sett just as dusk falls, and forage for food during the hours of darkness, returning from their wanderings before dawn. They are most active during spring and summer, but during the winter they are far more sedentary, spending much of the time underground in their sett, emerging only to forage during spells of milder weather.

Breeding takes place during early spring, but female badgers are able to delay implanting the fertilised eggs in their womb until later in the year, so that most cubs are born the following January or February. The cubs emerge in early spring, and are weaned during the summer months, leaving the sett later in the year.

There are currently more than 300,000 badgers living in the UK,

and that number is rapidly rising. But badgers face many threats to their survival: thousands are killed by motor vehicles every year, and it is now a common sight to see dead ones by the roadside in much of rural Britain. However, the greatest threat to the badger's continued survival is, without doubt, pressure from farmers to cull the animals.

The controversy over the badger's role in spreading bovine tuberculosis in cattle is an ongoing one, and there have been calls for wholesale culls as a result. However, badgers – and their setts – still remain protected by law. Ironically, some of the safest badgers are those that have managed to survive in our towns and cities as the urban sprawl has grown around them, finding havens in parks and wooded areas, and often feeding in suburban gardens.

Although the badger's diet is primarily earthworms, fruit and small mammals, they have also been implicated in the decline of hedgehogs, both in rural and suburban areas. Despite the hedgehog's spines, badgers seem adept at getting past this defence.

To see badgers, it is best to get in touch with an organised badger-watching group, as they will have the knowledge and expertise to take you to see a sett without disturbing its occupants. The experience is a memorable one: few other wild creatures have quite the same charisma as a family of badgers in a wooded glade on a summer's night.

Badgers forage for food from dusk to dawn, often coming into gardens to dig up lawns for earthworms.

Otter

The story of the otter is a heart-warming and inspiring one: one of our most iconic mammals disappeared from much of Britain and was on the verge of extinction until efforts by conservationists managed to turn its fortunes around.

Today, the otter has returned to every English county, and is also found in good numbers throughout Scotland, Wales and Northern Ireland. But only a few decades ago the situation was very different. A combination of hunting with specially bred otter hounds, persecution by anglers and the owners of fish farms, and, especially, the widespread pollution of our streams and rivers meant that the otter disappeared from many of its former haunts. The only places where it managed to survive were either upland habitats or on the coast – the maritime populations of northern and western Scotland were not so badly affected by either persecution or pollution.

Fortunately, in the nick of time, things changed. Otter hunting was made illegal in 1978, and during the same period a concerted effort was made to clean up some of our most polluted rivers. A fine example is the River Tyne, once considered to be Britain's most polluted river. Once the clean-up began, salmon returned upstream to breed, bringing with them a tiny population of otters. Since then, otters have become a regular sight – not just during the night hours, but often during the day as well. The same is true of many other British waterways, as well as

Few mammals are so well adapted to life in water or on land as the otter.

wetland areas such as the Somerset Levels and the Norfolk Broads, where otters are thriving.

The otter is one of the few native British mammals to be equally at home out of the water as in it. Unlike, for instance, seals, which are clumsy on land, otters are well suited to a terrestrial existence for part of their lives, and will sometimes travel from place to place by a land-based route.

But it is under the water that their grace really shows. They are transformed into a highly efficient killing machine, pursuing fish with a speed and agility that defies belief. Their special adaptations include their sleek, streamlined shape; short, dense fur, which traps air bubbles and provides insulation against the cold; and powerful webbed feet, which can propel them through the water at great speed.

Otters give birth to their litter of two or three cubs during the summer months, hidden away in a hole in a river bank, known as a holt. If you are really lucky, you may catch sight of cubs playing once they emerge from the holt, though the presence of otters is usually best detected by looking for signs such as footprints on muddy banks or their droppings, known as spraints, which have a distinctive sweet smell sometimes compared with newly mown hay!

Otherwise, you may be lucky to catch sight of an otter as it swims along a stream or river, or more likely on the coast, where they often feed among the seaweed at low tide. If you stay downwind, and remain quiet, you may be able to watch an otter as it goes about its daily business. But if you are looking for them inland, beware confusion with the closely related mink, an unwelcome escapee from North America, which is frequently found on the same waterways but is much smaller and darker in colour.

The otter's place in our affections was cemented by two popular works of literature: *Tarka the Otter*, by Henry Williamson (published in 1927), and *Ring of Bright Water*, by Gavin Maxwell (published in 1960). These both celebrated and demystified the otter, but today they are rarely read, which is a pity, as they provide great insights into the behaviour of this fascinating and beautiful mammal.

Seals

Two species of seal can be found off our coasts: the Atlantic grey seal and the smaller common seal, sometimes known as the harbour seal. Both are big and heavy creatures. Indeed, the grey seal is our largest regularly occurring mammal, with males sometimes approaching half a tonne in weight, comfortably heavier than our largest terrestrial mammal, the red deer. Both species spend the vast majority of their lives offshore, only coming onto land to breed, as their adaptations to an aquatic lifestyle make them clumsy on land.

Grey seals can be found around most of Britain's coastline, though they are generally absent from much of the south coast of England, and are most common in the north and west. Numbers have increased rapidly since seals were first protected in the nineteenth century, and the UK and Irish population estimate is now about 200,000 individuals – about half of the entire world population.

Despite its name, the colour of the grey seal is not the best way to tell it apart from its smaller relative. Instead, look for the grey seal's distinctive profile: its 'Roman nose' gives it a rather haughty appearance, compared with the common seal's friendlier, more dog-like muzzle.

Unlike most British mammals, which usually give birth in the spring or summer, grey seals do not come ashore to breed until the autumn. The females arrive first, and give birth to a single pup between September and December, depending on the location. Seal mothers are very devoted parents. They not only suckle the pup with high-protein and fat-rich milk, they also give up feeding themselves for the duration, sometimes for weeks on end. By the time the pup has shed its snow-white fur, it will have more than doubled in weight, from about 15 to 30 kg (33 to 66 lb).

The mother eventually abandons the pup, which remains on the beach for several days until hunger finally drives it to seek refuge in the sea. This is the most vulnerable time for baby seals: one autumn gale can sweep the pup out to sea before it is able to survive on its own. Meanwhile, the exhausted female has one more duty to

The common seal has a more friendly, dog-like appearance than its larger relative the grey seal.

perform. While she has been raising her pup, the males have been defending their patch of beach with a fierce determination. As with the red deer rut, the most successful male gets the lion's share of the females. Once the females are pregnant, they are able to delay implanting the egg into their womb for another couple of months. Combined with a gestation period of about nine months, this means the breeding cycle takes place at the same time every year.

The common seal, by contrast, breeds during the summer, mating one year and giving birth the following one. Just as with the grey seal, once the pup has been weaned, the female will mate again, so she is almost constantly pregnant. Despite their name, common seals are the scarcer of the two species, with fewer than 50,000 individuals around our coasts. They are mainly found along the east coast of England and Scotland, and the north and west coasts of Scotland, but rarely seen off southern or western England or Wales. In recent years, numbers appear to have dropped, causing concerns about the status of this charismatic marine mammal. The reasons for the decline are uncertain: pollution and fish shortages, together with disease, may be to blame.

By far the best way to see seals is to go on a specially organised boat trip. These run from a number of sites around the UK, notably north Norfolk (to the colony at Blakeney Point), southwest Wales, Cornwall, the Farne Islands off Northumberland, and western Scotland. Seals aren't the only attraction of these cruises. Look out for seabirds such as shearwaters, terns and auks, basking sharks (especially in the west during the summer months) and, if you are very lucky, whales, porpoises and dolphins.

The grey seal is Britain's largest breeding mammal.

Deer

The sight of red deer rutting on a misty autumn morning is one of the greatest of all British wildlife spectacles. There is something about the roar of the stags, as they lock antlers in an all-or-nothing battle to win the favour of the watching hinds, that sends shivers up the spine. The top males have a tougher task each year to defend their position as the dominant stags, because a new generation of males is always ready to challenge them for supremacy.

Key elements for success in the rut are the size and complexity of the antlers – the greater the number of 'points', the more dominant the stag – and the position at the centre of the rutting arena. The females stand by as if they are mere spectators but, as with all such courtship displays, they in fact play the key role, by selecting the winning stag as their mate. In this way, the female guarantees that her offspring will have the best genetic heritage.

But the annual rut is just one aspect of the lifecycle of these majestic animals, Britain's largest terrestrial mammal. Away from this battleground, they live a quieter life, in small herds of males or females, feeding on leaves, grasses and, during winter, the bark of trees, before the rutting season begins again.

Red deer are the largest and most majestic of Britain's six species of deer, although if you count the feral herd of reindeer on the top of the Cairngorms, they are pushed into second place. Of the other species, only the much smaller roe deer is, like the red deer, native to our shores, but even these have had to have a helping hand: the English population became extinct in Victorian times and was reintroduced there at a later date.

Today, roe deer are thriving in much of the British countryside and even in some green areas of our town centres, though because they are generally solitary or in pairs, rather than herds, they are not as noticeable as some other species. Unlike red deer, male roe deer have two short, pointed antlers sticking up from their crown.

The other deer often found in deer parks, and which also takes part in an annual rut each autumn, is the fallow deer. The Normans brought this species here, and so it has been a British resident for almost a millennium. Fallow deer are sociable creatures and can often be seen in large herds. They are commoner in the south of Britain, with strongholds being the New Forest and London's Richmond Park.

A red deer stag, showing off his magnificent set of antlers.

The three other kinds of deer found in the wild in Britain – sika, muntjac and Chinese water deer – all originate from Asia, and were brought here in the nineteenth and early twentieth centuries to adorn the parks surrounding stately homes such as Woburn, north of London.

Unfortunately, they escaped and have since spread, with varying degrees of success. Sika deer are mainly found in north and west Scotland, where they have interbred with native red deer; Chinese water deer are pretty much confined to a few locations in East Anglia; while the smallest species, the muntjac, has spread rapidly northwards and westwards across most of England. This diminutive creature, about the size of a medium-sized dog, is now a major problem, as it destroys much of the woodland understorey, causing problems for nesting birds such as the nightingale.

For good views of deer, the best places to visit are deer parks such as Fountains Abbey in Yorkshire, or Richmond and Bushy Parks in southwest London, where semi-wild animals will often allow a close approach, giving great opportunity for photographs. But be very careful during the rutting season, when these normally placid animals often turn aggressive, and can cause injury if you get too near.

Roe deer are increasingly common and easy to see, especially in southern and western England.

Fallow deer were brought to Britain after the Norman Conquest, and have thrived in our woodlands and parks.

Roe deer are generally seen alone or in pairs rather than in large herds.

Rabbit & Hares

With their long legs and ears and lolloping gait, these three long-eared mammals – the rabbit and the brown and mountain hares – are superficially very similar and yet their lifestyles are very different indeed. Only the rabbit digs burrows to hide from predators, while the two hare species rely on camouflage, cunning and, as a last resort, speed, to avoid being caught and eaten.

Our attitude towards these creatures is also rather different. Rabbits were probably brought here by the Romans (and, certainly later on, by the Normans), as a convenient source of fur and meat, yet, despite their long association with us, they are usually regarded both as a foreigner and a pest. This is perhaps because they are more likely to invade our gardens and nibble the lettuces, or make burrows. The brown hare is also, technically, an alien species, though it has been here even longer than the rabbit, possibly since the Iron Age.

Along with the grey squirrel, the rabbit is the most successful of all introduced non-native mammals.

Whatever its origins, though, we tend to treat the brown hare with far more reverence, regarding it as a rather magical creature, associated with all sorts of folklore and beliefs. This derives partly from the hare's apparent ability to disappear when being chased. Instead of doing what rabbits do and diving down a burrow, hares simply lie still in a shallow depression in a field, known as a form. This made the hare appear to vanish, which perhaps led to the species being granted magical qualities.

Seen well, it is easy to tell a hare apart from a rabbit. The latter is a small, stocky animal, with relatively short hind legs and ears, and a white 'powder puff' of a tail, clearly visible as it runs away. The hare is a much larger and more elegant creature, with long ears, body and hind legs, enabling it to run at almost twice the speed of the fastest human sprinter. When seen from behind, they have a dark centre to the tail, a useful identification feature.

The breeding behaviour of rabbits and hares differs in several important aspects too. Rabbits have several litters every year, in a series of burrows, known as a warren. Each litter can number up to seven young, called kittens. Although they are born hairless and blind, within as little as four months the young can breed for themselves. Hares also have several litters of

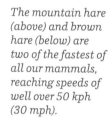

young (leverets) each year, but they are raised above ground instead of in the safety of a burrow.

Rabbit populations have often fluctuated, especially because of the effects of the disease myxomatosis, which wiped out 95 per cent of Britain's rabbits during the middle decades of the twentieth century, and from which the population took some time to recover. But brown hare numbers have stayed more stable. Today, there are fewer than one million hares, but over 35 million rabbits – roughly one for every two human beings.

The brown hare's sibling species, the mountain hare, is our only native member of the trio. It is one of only three British creatures (the other two being the ptarmigan and stoat) that turn white in winter. For those living in the far north, in the snow-covered high tops of the Cairngorms, this makes perfect sense, but farther south and west, mountain hare populations in Ireland (known as the Irish hare) stay brown all year round, as there is little or no snow. The mountain hare's main predators are golden eagles and foxes, but stoats may also hunt them – a rare example of one camouflaged animal pursuing another.

Of all our mammals, perhaps only the fox, the red squirrel and the otter have featured so often in our children's literary heritage as the rabbit. Books such as Richard Adams' *Watership Down* and Beatrix Potter's Peter Rabbit stories, have delighted generations of children, and do so to this day.

The mountain hare (above) and brown hare (below) are two of the fastest of all our mammals, reaching speeds of well over 50 kph (30 mph).

Stoat & Weasel

These two small and elusive predators are, gram for gram, by far the fiercest of all our carnivores. Watching a stoat catch and kill a rabbit weighing five times as much as it does is a truly awesome sight, while the world's smallest carnivore, the weasel, is equally effective at dispatching small mammals with a single bite to the back of the neck.

Stoats and weasels are often confused with one another, though, given good views, they are reasonably easy to tell apart. Stoats are considerably larger – up to four times the weight of a weasel – and have a very distinctive black tip to their tail, often the last thing you see as the animal dives into dense cover. The weasel's tail is also generally shorter and less bushy than that of the stoat. Otherwise, the two are superficially similar: both are long, slender creatures with chestnut-brown upper parts and pale, often almost yellow, under parts.

Both are solitary hunters, usually doing so by night or at dawn and dusk, when their prey is also likely to be more active. They rely on excellent eyesight, a good sense of smell and a rapid turn of speed. Once they have the opportunity, they can be ruthless killers, using their incredibly sharp teeth to bite the neck of their quarry, then dragging the corpse out of sight so they can feed at leisure.

Both stoats and weasels are found throughout England, Scotland and Wales, though the weasel is absent from Ireland, suggesting that it recolonised Britain after the last Ice Age at a later date than its cousin, after the land bridge between Britain and Ireland was covered by the rising Irish Sea.

In the northern parts of the stoat's range, especially at high altitudes where snow is most frequent and heavy, they adopt an all-white winter plumage, known as ermine. Just one feature – the tip of the stoat's tail – remains its original black, a feature that can be seen in illustrations of medieval royalty wearing white ermine cloaks with dark spots.

Stoats generally have just one litter a year, compared with the weasel's one or two, but they make up for it with the number of young: as many as a dozen compared with the weasel's four or five. The young kits can be very playful, chasing each other around madly in mock-fights, which comes in useful later in life when they must learn to hunt for themselves. This happens sooner

Weasels may be small, but they make up for their tiny size with a ruthless ability to kill their prey.

than you might think: young stoats and weasels leave their parents to fend for themselves after less than three months.

Both stoats and weasels feed on a range of prey, including voles, mice and shrews as well as rabbits. Occasionally, they also take baby birds and eggs from nests, as well as fruit and berries in autumn. In turn, they may fall victim to attack by foxes, owls and other birds of prey such as the kestrel, as well as domestic cats, especially in suburban areas. Like other carnivores, they have long been regarded as vermin by gamekeepers and farmers, and have been ruthlessly trapped, shot and poisoned as a result. However, this persecution may now be on the decline, especially as many stoats and weasels now live in parks and woods on the edges of towns, rather than in the open countryside.

Both species are doing well in Britain, with roughly half a million individuals of each, a steady increase over the past few decades. Away from the British Isles, they are highly successful too: both these adaptable little predators are found right across the northern hemisphere, in Europe and North America well beyond the Arctic Circle, where, like the stoat, the weasel also turns white in winter. The weasel is also found as far south as North Africa.

Stoats are naturally curious creatures and are easier to see than their smaller relative.

Bats

Of Britain's fifty or so species of non-marine mammal, about one in three are bats. Yet of all our mammals, bats remain among the most elusive, perhaps because of their nocturnal habits and an indefinable air of mystery surrounding them.

Tales of bloodsucking vampires have led many people to fear bats, even though our native species are harmless. Indeed, because they use the technique of echolocation to find their prey and avoid bumping into things, there is no likelihood of getting a bat caught up in your hair – a common but entirely misguided fear. Nor are bats blind, as is often assumed, and nor do they generally carry or transmit rabies, though if handled carelessly, they can give you a nasty bite.

The truth about our bats is even more amazing than the myths and folk tales. Bats are the world's only truly flying mammals (so-called 'flying squirrels' merely glide), and are consummate night hunters: a single bat can catch and eat up to 3,000 insects in one night. The skill with which a bat pursues its prey, zeroing in on the target before lifting its hind claws to grab the creature out of the air, is quite awesome.

Of course, actually witnessing such an expert act of predation is not easy. Getting to know bats at all requires patience and a willingness to go out after dark. Known roosts are a great place to start. At dusk, you can watch hundreds, sometimes thousands, of bats pouring out of cracks and crevices in buildings and taking to the wing, as they begin their night's hunting.

But if you are really serious about bats, you will need to get a bat detector: a portable piece of electronic wizardry that can read the bats' high-pitched sounds (too high, usually, for us to hear), and then tunes in to the right frequency. Once you do so, the bat's inaudible sounds will suddenly become audible, as a series of clicks, grunts and wheezes. With expert help and experience, you can then identify the bat you are 'hearing' by its frequency and rhythm.

Many of Britain's seventeen or so species of bat are both very rare and very hard to find. These include the greater and lesser horseshoe bats, both of which are confined to southwest England and Wales; Bechstein's bat, which depends on ancient woodland for its home; and Leisler's bat. The greater mouse-eared bat has only a toehold as a British breeding species and is almost certainly now extinct here.

However, several species of bat are still widespread and common. These include our largest bat, the noctule, which often appears in

the sky an hour or so before sunset, giving excellent views. Other reasonably widespread bats include Daubenton's, which is often found dipping down over stretches of water to feed, rather like swallows or martins; and the brown long-eared bat, which, as its name suggests, has long, rabbit-like ears. Even so, some of these species have been in decline, partly because agricultural pesticides have reduced their insect food, and also because new buildings have reduced the availability of nest sites under eaves.

Of all our native bats, by far the commonest and most widespread are the smallest and lightest, known as pipistrelles. In recent years, though, bat watchers have discovered that the 'common pipistrelle' is in fact two quite separate species, with different habits and populations. The key distinction between them is that the common pipistrelle echolocates at a frequency of about 45 kilohertz, while its almost identical cousin, the soprano pipistrelle, does so at a much higher frequency, about 55 kilohertz. This keeps the individuals of both types apart, meaning that they are definitely different species. This amazing fact makes many bat watchers wonder what else there is to be discovered about these enigmatic and fascinating creatures.

Long-eared bats have particularly well-developed echolocation in order to hunt down flying insects at night.

Small Mammals

An army of small mammals – rats, mice, voles and shrews – live their lives almost within touching distance of us but we hardly ever see them. Apart from the odd encounter with a house mouse, or occasionally a brown rat, these creatures are mostly invisible. And yet they are among the commonest wild animals in Britain. Indeed, the most abundant, the field vole, is the only wild mammal that outnumbers the human inhabitants of these islands.

The reason we hardly ever see them is partly because many lead nocturnal lives, but it is perhaps more to do with the fact that they are extremely vulnerable to attack by predators, and so tend to hide out of sight. Small mammals have an array of predators, ranging from foxes and badgers, through kestrels and buzzards, to those consummate nocturnal hunters the barn and tawny owls. It is no wonder that these creatures spend so much of their lives hidden away.

Yet small mammals cannot hide all the time because they need to eat, and in vast quantities. The smallest species, such as common and pygmy shrews, must consume close to their own body weight every day if they are to survive, which means they must be on the go virtually all the time. This need to replenish lost energy may make them very vulnerable to attack.

Small mammals fall into two categories: rodents and insectivores. Rodents include rats (two British species), mice (four species) and voles (four species), as well as squirrels and dormice, while the four species of shrew found in Britain are insectivores, a group that also includes the mole and the hedgehog. The word 'rodent' derives from a Latin term meaning 'to gnaw', and all rodents have sharp front teeth, which are continually growing as they wear down.

Rats and mice are probably the least popular British mammals, apart from, perhaps, the introduced grey squirrel. Yet we should admire them for their adaptability, especially the brown rat and house mouse, which have learned to take advantage of

All rodents have extremely sharp teeth, as they are constantly feeding to maintain their energy levels.

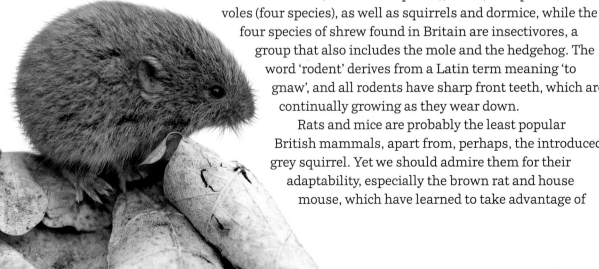

our wasteful habits in order to feed and survive. A second species of rat, the black, or ship, rat, was once common in ports and dockyards but now has the distinction of being Britain's rarest mammal, confined to the Shiant Isles, off the west coast of Scotland.

Two other rodents – the wood mouse and yellow-necked mouse – live in more 'natural' habitats than the house mouse, although they, too, have had to adapt to a rural landscape increasingly dominated by farming. Wood and yellow-necked mice have also adapted to live near people, often visiting garden bird tables in order to grab morsels of seeds or nuts. They are hard to tell apart, but both look very different from the house mouse, having larger ears and eyes, which makes them look rather endearing.

Britain's shrews are curious little creatures. They include our smallest terrestrial mammal, the pygmy shrew, which weighs in at just 4 g (⅛ oz) or so, and our only venomous one, the water shrew, which can inject paralysing saliva into its aquatic prey. The most abundant and widespread is the common shrew, whose UK population numbers in the tens of millions, but, like its relatives, it is hardly ever seen. By far the rarest British shrew is the Scilly (or lesser white-toothed) shrew, a continental European species that in Britain is confined to the Isles of Scilly. Given its widespread distribution in Europe, it is likely that the Scilly shrew was brought to the islands by passing traders or sailors.

Another island creature probably brought here by accident is the Orkney vole, a unique race of the European common vole found only on this northern archipelago. By far the commonest vole in Britain is the field (sometimes known as the short-tailed) vole. This unremarkable little creature nevertheless plays a vital role in our ecosystem, providing a large part of the diet of so many of our predatory birds and mammals. There are thought to be about 75 million field voles in Britain, making it our commonest mammal by some measure.

Mice, voles and shrews are rarely seen, as they spend their time avoiding aerial predators such as kestrels and owls.

Water Vole

The water vole has an unenviable distinction: it has declined more severely in recent years than any other British mammal. This is due to a number of factors, including the widespread pollution of water in our rivers and streams by farm chemicals and other industrial poisons during the latter years of the twentieth century, which killed off much of the population of this aquatic creature.

But by far the greatest problem facing the water vole has been the spread of an alien creature: the North American mink, one of the water vole's most effective and ruthless predators. Mink were brought here for fur farming, and then either escaped or were deliberately released by foolish and misguided animal rights campaigners, only to wreak havoc on one of our favourite native mammals. During the final decade of the twentieth century, water vole numbers fell by 90 per cent, and today the creature is still absent from many of its former strongholds.

There is some good news, though. In recent years, the comeback of another aquatic mammal, the otter, has helped drive mink away from some waterways. Otters still eat water voles – as does a wide range of other creatures, including herons, foxes, stoats and even pike – but they do not seem to cause quite such devastation as the mink have in such a short time. Meanwhile, the cleanup of our rivers, streams and canals has also given water voles a helping hand.

Seen well, a water vole is unmistakable, but if you only have brief views, they can look surprisingly similar to a less endearing mammal, the brown rat. They are our largest vole, about 15–20 cm (6–8 in) long, with another 7–10 cm (3–4 in) of tail, and weighing about 200–300 g (7–10 oz). Their shape best identifies them: dumpy, with small ears, a squashed snout (not a long, pointed one) and a short, furry tail. Colour varies, depending on the light and how wet the animal is, but they usually appear a blackish-brown.

If you see what looks like a small version of the water vole, you may well be looking at its much more compact relative, the bank vole. Less than half the length of its larger cousin, and weighing just 40 g (about 1½ oz), the bank vole is much commoner than the water vole but just as seldom seen, as they usually live in dense vegetation to avoid predators.

Water voles are mainly vegetarian, feeding mostly on grass, sedges and reeds. They raise their families in burrows dug and have several litters each year; their young are born blind and helpless, but soon grow and leave the safety of their burrow.

If you want to see water voles, they can still be found on slow-moving bodies of water all over Britain, apart from the far north of Scotland and many offshore islands. Look out for closely cropped grass, known as lawns, outside their small, neat burrows, which they make along the banks. You may be lucky enough to see a vole's head poking above the surface of the water as it feeds or swims along a river, stream or canal. Surprisingly, perhaps, they are often much easier to see in urban and suburban settings, as they get used to disturbance from dog-walkers, joggers and other passers-by, and so are more tolerant of the watching naturalist.

No account of the water vole would be complete without a reference to its most celebrated appearance in our culture: as 'Ratty' in Kenneth Grahame's children's novel *The Wind in the Willows*. The deserved success of this book, together with numerous TV, radio and stage versions, means that we are all aware of water voles. What a pity, though, that Grahame chose to call his character a 'water rat', thus fostering the confusion between water voles and brown rats, which continues to this day.

One of our most charismatic yet fastest-declining mammals, water voles can still be found at wetland sites throughout most of Britain.

Dormouse

With its long, bushy tail, the edible dormouse looks more like a miniature grey squirrel than a mouse.

The dormouse has two claims to fame: it hibernates for longer than any other native British mammal and it featured in arguably the best-known children's book of all time, *Alice's Adventures in Wonderland*, where the aforementioned rodent sleeps soundly through a surreal tea party featuring Alice, the Hatter and the March Hare. It is also the only British rodent with a furry tail, which, together with its honey-coloured fur, small ears and large eyes, gives it a far more cute and endearing appearance than our other rodents.

The name 'dormouse' derives from the French verb *dormir*, meaning to sleep. The more widespread of the two British kinds is the common, or hazel, dormouse, one of about 30 species in its family, spread across Europe, Africa and Asia. Dormice are mainly found in southern Britain, usually in deciduous woodland, though they can also live in conifer plantations. They are rarely seen, though, primarily because, unlike other rodents, which live a largely terrestrial life, dormice prefer to live in the woodland canopy.

They do so because of their diet. In spring and summer, they subsist almost entirely on flowers and pollen, and occasionally aphids, while during the early autumn, they switch to feed on berries, nuts and other fruits. This enables them to build up enough fat reserves to see them through the long period of hibernation, generally six months or more, from October through to April, even May, if weather conditions are poor. Dormice even occasionally go into a state of torpor, very similar to full hibernation, during the spring and summer; either because of food shortages or to avoid bad weather.

Dormice only come down from their treetop homes in autumn, when they make a winter nest beneath the leaf litter of the forest floor; they also readily take to specially provided nest boxes or even tennis balls, which have been cut open to allow them to enter! Nest boxes are not just convenient for the dormice; they also allow naturalists to monitor their populations and study the behaviour of this otherwise elusive creature. Another way to discover whether there are dormice living in a wood is to look for discarded nut shells – unlike mice or squirrels, dormice teeth leave a very neat hole with a smooth, bevelled rim.

Despite being asleep for so much of the year, dormice have a frantic breeding cycle. The first litter of around four or five young can be born as early as May but generally appear later in the summer. If the weather stays warm into September and October, they may sometimes give birth to a second litter, though these may struggle to survive if there is a sudden cold snap soon afterwards. The young are, like most small mammals, born hairless and blind, but once they mature, they can live as long as five years.

A second species, the edible dormouse, was introduced to Hertfordshire, just north of London, at the very beginning of the twentieth century, by the pioneering conservationist Walter Rothschild. They haven't moved far since, and remain within a radius of about 30 km (20 miles) from their original home. Looking more like a miniature version of a grey squirrel than a mouse, they may appear cute, but they can cause a lot of damage to property, as they sometimes live communally in attics. The Romans considered the edible dormouse to be a great delicacy, and served it in vast numbers at their famous feasts, sometimes dipped in honey as a rather unusual dessert.

Edible dormice are mostly nocturnal, and spend much of their time in specially made nests in woodlands or inside the roofs of our homes. They hibernate for just as long as their smaller cousin: from early autumn right the way through to the following April or May.

With its huge eyes, few other British rodents are quite so endearing as the dormouse.

Harvest Mouse

Few other mammals have adapted quite so well to living on arable farmland as the harvest mouse. This tiny rodent, about 8 cm (3 in) long and weighing as little as 4–5 g (about ⅙ oz), has special adaptations that enable it to live among fields of crops, where it uses its prehensile tail to grip onto grass stems, like a miniature monkey.

Of Britain's mammals, only the pygmy shrew and the pipistrelle bat are smaller. Indeed, the harvest mouse is Europe's smallest species of rodent, well deserving of its scientific name *Micromys minutus*, which roughly translates as 'tiny little mouse'.

Harvest mice are found right across the temperate regions of Europe and Asia, as far to the east as China and Japan, though they are scarcer the farther north you go. In Britain, they have a southerly distribution, being found mainly south and east of a line between the Severn and the Humber estuaries.

Due to their retiring habits, harvest mice are rarely seen, but if you do catch sight of one, its russet-brown coat, white belly, blunt snout and tiny ears are distinctive. Like other rodents, they are mainly active either after dark, or at dawn and dusk, though they can also sometimes be seen during the day. Despite their small size, which means that retaining body heat can be a problem in cold weather, they do not hibernate in winter. However, like other small mammals, they are less active at this time of year.

Although they are usually linked with the classic farmland crops of wheat and oats, the harvest mouse is equally at home in long grass, hedgerows and even reed beds, where they weave their grassy nest between the stems and use their long, prehensile tail to avoid falling into the water beneath.

Another adaptation to life in a vertical world of grass stems and reeds is the harvest mouse's broad feet, which have a much larger big toe than other rodents. This enables it to grip stems or reeds with its hind feet and tail, leaving the two front paws available for grabbing food. The harvest mouse's diet mainly consists of grains and seeds, but they also eat fruit and berries, and the occasional insect. During the autumn and winter months, they store food so that when a cold snap hits, they are still able to survive. Their usual lifespan is about 18 months.

The harvest mouse's nest is such a characteristic structure that it was originally discovered before the mammal itself was described.

In the pages of his famous work *The Natural History of Selborne*, the eighteenth-century naturalist Gilbert White describes seeing this extraordinary object, woven neatly from stems of wheat, 'about the size of a cricket-ball', and containing eight naked, blind baby mice.

The female harvest mouse is pregnant for about two-and-a-half weeks, after which she gives birth to a litter of seven or eight young, known as pups. Two weeks later, the pups have been weaned, and soon afterwards the parents abandon them, though they may still return to the safety of the nest. Not long after that, the female may become pregnant again, as she can give birth to as many as three litters in a single summer season.

Like many of our more specialised mammal species, especially those that live on farmland, harvest mice have declined in the years since World War II. The decline has occurred as a result of modern farming techniques, which remove the hedgerows that give the mice shelter. Modern combine harvesters can also be lethal to harvest mice, as they give them no chance of escaping when the crop is being harvested. That said, there are still more than one million individuals in Britain.

The harvest mouse uses its prehensile tail to cling onto grass stems when feeding.

Hedgehog

Few British mammals are thought of with such affection as the hedgehog. This spiny insectivore has endeared itself to generations of gardeners, thanks to its preference for feeding on slugs, while householders love it for its approachability – not many other mammals will come to feed at our back doors.

Yet this delightful animal is now in big trouble. Hedgehogs have been in decline for several decades now, due to a combination of loss of habitat, predation, poisoning and being run over by motor vehicles. One problem for the hedgehog occurs in spring, when it needs to roam to find a mate. In urban and suburban areas, in particular, this usually entails crossing busy roads, putting it in the firing line of car tyres. Nevertheless, there are still approximately one million hedgehogs living in Britain, everywhere apart from remote offshore islands and the high mountain ranges of the north and west.

As our only spiny mammal, the hedgehog is unmistakable. The spines on its head and back are incredibly closely packed: there are more than 5,000 of them, providing an excellent defence against danger. When a predator approaches, the hedgehog simply curls itself into a tight ball, raises its spines and thus avoids being killed and eaten.

Unfortunately for the hedgehog, a badger's claws are sharp and strong enough to break down this defence, with the result that these

Hedgehogs prefer hard winters as this enables them to hibernate properly without being tempted to emerge too early.

large predators kill many hedgehogs. In such circumstances, it is often better for the hedgehog to run – for such a small animal they can move very fast when they need to.

Hedgehogs have a varied diet, feeding on a wide range of invertebrates, especially earthworms, slugs, beetles and caterpillars. This presents a dilemma for the gardener: using slug pellets can kill hedgehogs, so it is far better to let this natural pest controller do its job without the danger from poisoning.

They will also scavenge for carrion, and take the eggs of ground-nesting birds. This became a huge problem when some hedgehogs were released onto the islands of North and South Uist and Benbecula, in the Outer Hebrides, off the northwest coast of Scotland. These islands are home to thousands of pairs of nesting waders, including some of the world's densest breeding populations of species such as the dunlin, redshank, oystercatcher and ringed plover. The hedgehogs wreaked havoc, and seriously threatened the waders until a project began to capture them. Culling the animals seemed the only option, until a campaign led to a change of heart. Captured hedgehogs are now transferred to the mainland, thus not only ridding the islands of a problem but also helping maintain the rest of the British hedgehog population.

The hedgehog's lifecycle is based around breeding and hibernation. The female gives birth to about half a dozen young during the late spring or summer, in a nest on the woodland floor or in a corner of a garden. The young stay dependent on their parents for about a month, and then begin their own wanderings. Sadly, their chances of survival are not high. As well as predation and starvation, hedgehogs often fall victim to problems during hibernation. They need to be careful where they choose to spend the winter: some 'log piles' turn out to be bonfires, which means that the unfortunate animal is incinerated come 5 November.

Even if they do survive Guy Fawkes' Night, hibernating hedgehogs face another problem. During mild winters, the temperature often rises high enough to fool them into emerging well before the winter is truly over. But it will then not be able to find enough food to maintain its critical weight until the spring. As a result, many die prematurely. Hard winters are in fact much better for hedgehogs, as they enable the animals to hibernate properly until March or April, when they emerge.

Pine Marten & Polecat

These two relatives of the much more widespread stoat and weasel are among our most handsome yet elusive wild mammals, rarely seen unless you have special access to one of their regular haunts.

Of the two, the pine marten is primarily found in the north and west, with its main range being Highland Scotland and parts of Ireland, with just a handful of animals scattered across the wilder parts of England and Wales. Wales, conversely, is the stronghold of the polecat, which can also be found in parts of central and northwest England, and a few locations in Scotland. The polecat is by far the commoner of the two species, with perhaps 40,000 individuals in all, compared with fewer than 4,000 pine martens.

Nevertheless, both can be equally tricky to see, being wary of humans (for good reason, as they were persecuted for centuries), mainly nocturnal and extremely secretive in their habits. If you do manage to get a good view, they are pretty straightforward to identify. The pine marten looks like a large stoat, with a long nose, rounded ears, dark chocolate, almost chestnut upper parts and creamy yellowish under parts. The polecat is smaller and slimmer, and more uniform in colour, with pale hairs beneath dark ones, so that the shade changes depending on the light and angle of view.

The pine marten is a shy and elusive forest-dweller, found mainly in the Scottish Highlands.

Both animals are agile predators, feeding mainly on rodents such as mice and voles, but also invertebrates, berries and birds' eggs. One pine marten near Loch Ness regularly came to a garden bird table, showing a particular liking for jam sandwiches! Neither species has many natural predators, though foxes and, in the case of the pine marten, golden eagles may occasionally take them, as may domestic dogs.

Pine martens breed during the summer, usually hidden away in a dense coniferous forest. They have two young, known as kits, which remain in the safety of the den for several months. Although they can live more than ten years, they do not usually breed until they are three years old. Polecats have much larger litters – between five and ten kits in all – and these stay with the parents until they are two or three months old.

Outside the UK, pine martens are found throughout the wooded, temperate regions of Europe and parts of western and southwest Asia. They share much of this range with another forest-dwelling relative, the beech marten. The polecat's world range is similar, though it does extend farther east into Asia.

Without doubt, the biggest threat to these species, especially to the more common and widespread polecat, has been human beings – it was almost exterminated as a British breeding mammal during the nineteenth and twentieth centuries. However, in recent decades, persecution has declined, and the species is making a comeback, both in numbers and by extending its range out of its remote strongholds and into more populated areas. Here, ironically, the main threat to their survival may be their hybridisation with domestic ferrets.

The pine marten, too, suffered from heavy persecution from gamekeepers and landowners, especially in the forests of the Scottish Highlands, and it also declined due to widespread loss of woodland habitat during the same period. Like the polecat, the pine marten has made a recovery in recent years, but it remains one of our rarest and most elusive mammals. One benefit of the rising population is that pine martens have been discovered to be a very effective predator of grey squirrels, helping to limit the range of extension of this species into parts of Scotland.

If the polecat looks familiar, that is because it is the ancestor of the widely domesticated ferret. These creatures have been used by humans for hunting rabbits and other quarry for centuries, possibly since pre-Roman times. They were especially popular in the north of England during the century or so following the Industrial Revolution, but they are not kept as much nowadays, possibly because they are known to have a nasty bite.

Polecats have more than a passing resemblance to their cousin the domestic ferret.

Red & Grey Squirrels

The British are known for their patriotism, even when it comes to wild animals. Our attitude to the two species of squirrel found in the wild in Britain – the native red and the introduced grey – is the perfect example. The red is held up as a paragon of virtue, the innocent victim of the evil grey. The fact that the grey is indeed an alien species, brought here in the late nineteenth century from its native North America, does appear to justify this distinction. Some have even suggested that the grey squirrel allows us to give vent to anti-American feelings under the guise of concern for the red.

As a result, all sorts of myths abound regarding the two species, one being that the larger and more aggressive grey squirrels drive the reds from their homes in a form of 'ethnic cleansing'. The truth is more complicated: grey squirrels carry a virus, known as squirrelpox, which does not harm them but is lethal to the red squirrel. Grey squirrels cause problems for forestry managers, too, as they strip the bark from trees, and they also take the eggs and chicks of woodland birds. Nevertheless, it should be remembered that the red squirrel was also once considered to be a pest.

In the century or more since the grey squirrels were brought to southern Britain, they have extended their range northwards, and at the same time the red's range has retreated. Today, red squirrels are mainly confined to northern and western Britain, with their stronghold being the Highland forests of Scotland. Small relict populations continue to hang on in parts of Wales, Northern Ireland, the Lake District, the Lancashire coast, and two south-coast islands that the grey squirrel has not managed to colonise: Brownsea Island and the Isle of Wight. In contrast, grey squirrels are now widespread throughout England and Wales, and are rapidly heading farther north into the Highlands of Scotland.

To prevent them reaching the red squirrel's heartland, the government

Grey squirrels (below) and red squirrels (opposite) can rarely coexist in the same habitat; inevitably the stronger grey usually displaces the smaller red.

has recently announced that grey squirrels will be culled in parts of their range. A nationwide cull has sometimes been suggested, but with at least 2.5 million individuals, many of them living in our towns and cities, this would be impractical, if not impossible.

Even though it is unfortunate that this non-native species has been allowed to colonise Britain in such an unfettered way, we should also recognise that for many of our nation's children – especially those living in urban areas – the grey squirrel is the only wild mammal they will regularly see. The antics of the greys as they try to reach the food in hanging bird feeders also provide plenty of entertainment; indeed, the BBC has regularly featured the acrobatics of grey squirrels as they tackle man-made obstacle courses, to the enjoyment of millions of viewers.

Telling the two species apart should be easy, yet some individuals can cause confusion. Reds are usually a bright reddish-brown colour with neat white underparts and pointed ears, while greys tend to be mainly grey with more rounded ears. But some reds can look very dark, and greys can be surprisingly brown. To confuse the matter still further, one population of grey squirrels in Cambridgeshire and Hertfordshire is made up of virtually black individuals. These appear at first sight to be a separate species but they are in fact a genetic variation of grey squirrels that originally came from the area around Niagara Falls in Canada.

Like many mammals, both red and grey squirrels are less active during the winter months, though they do not actually hibernate. Both build nests, known as dreys, from twigs, leaves and moss, and breed twice a year: the red during late winter and midsummer; the grey in late winter and autumn. Both also store nuts – especially hazelnuts and acorns – during the autumn, which they dig up and eat later in the year when food becomes scarce.

Mole

Like the hedgehog and shrews, the mole is an insectivore but, unlike them, it lives primarily underground. Were it not for the regular appearance of mounds of earth, known as molehills, most of us would be unaware of its existence. Despite its abundance – more than 30 million moles live in England, Scotland and Wales (the species is absent from Ireland) – moles are very rarely seen.

No other British mammal is so well adapted for a subterranean existence. About 15 cm (6 in) long, and weighing about 100 g (3½ oz), the mole is a slender creature with small ears, virtually non-existent eyes, a pointed snout and two huge forelimbs that enable it to dig at an incredible rate. These claws benefit from an extra 'thumb' – actually, a modified bone from the animal's wrist – which gives them even more digging power, so they can shift considerable amounts of earth as they dig their tunnels. Indeed, the name mole is a contraction of the Middle English 'mouldiwarp', which means 'dirt-thrower'.

Moles have two other special adaptations. They are able to tolerate higher levels of carbon dioxide in their blood than other mammals, which allows them to spend more time underground in this low-oxygen environment. In addition, their fur is able to lie both ways, so that the animal can move both forwards and backwards along its network of tunnels without hindrance. This is especially useful if the mole needs to reverse rapidly to avoid a predator or sudden flooding. This characteristically smooth fur meant that before the arrival of mass-produced, man-made fabrics, moleskin was a valuable commodity for making waistcoats and other items of men's clothing.

The mole's most extraordinary achievement, though one that is rarely appreciated by Britain's gardeners, is its ability to dig extensive networks of underground tunnels, often many metres long. These enable the mole to avoid being caught and eaten by predators, and also provide it with food: worms, slugs and other invertebrates will simply fall into the hole linking the tunnel to the open air, providing the constantly patrolling creature with a regular supply of things to eat.

Moles also have a toxin in their saliva that paralyses their prey. They can then store their still-living victims in underground larders, which sometimes contain thousands of earthworms.

Their underground lifestyle means that, unlike most other small mammals, their lifecycle is not governed by the usual diurnal and nocturnal rhythms. Instead, moles work a 'shift pattern' of a few hours

of activity followed by a few hours' rest, but they must feed regularly when they are active because, like other insectivores, they need constantly to replace lost energy. Contrary to popular belief, however, they are not blind. Despite spending the vast majority of their life in darkness, they can see through those tiny eyes.

Moles are generally solitary but will meet up to breed during late winter and early spring. The female (sow) gives birth to a litter of three or four pups in April or May, which she continues to feed and look after for another month or so. Occasionally, the sow will have another litter later in the year. The youngsters are particularly vulnerable as they disperse to find new homes. Many are killed on the road, and their predators include cats, tawny owls, stoats and buzzards. If they do survive, and manage to find a suitable place to dig their own network of tunnels, they may live for as long as five or six years.

Once moles were hunted down by professional mole-catchers, who made a tidy living selling the skins, but nowadays they are more likely to be killed by pest controllers. Poisoning using strychnine has now been banned, partly because it killed many other creatures as well as the target species. But, farmers and gardeners apart, the British have a great affection for moles, not least because, as with the water vole and badger, the mole in the children's novel *The Wind in the Willows* by Kenneth Grahame is such an endearing character.

The mole is the only British mammal adapted to a life spent almost entirely underground.

Whales & Dolphins

When we think of Britain's wildlife, it's easy to forget that we are surrounded by incredibly productive seas, which, although often inaccessible to us, are home to some of the world's most spectacular creatures: whales, dolphins and porpoises, collectively known as cetaceans.

Almost 30 different species of cetacean have been seen, at one time or another, off the coasts of Britain. These include the largest animals on the planet: blue, fin and sperm whales, together with the largest marine predator, the killer whale, or orca, and some of the smaller sea mammals, the dolphins and porpoises.

Of all our marine mammals, the easiest to see are the harbour porpoise and bottlenose dolphin. Both live around our coasts, mainly in inshore waters, and both frequently approach boats, showing the curiosity for which they are famed. The harbour porpoise is the smaller of the two, about 1½ m (5 ft) in length, compared to more than 3 m (10 ft) for the dolphin. They are best identified by their swimming action: when they surface, they appear to roll through the water. The shape of their dorsal fin (the one along their back) is distinctive too, small and triangular in shape, compared to the dolphin's longer and more curved and pointed fin. Harbour porpoises often come close to shore in sheltered bays, either on their own or in small groups.

Dolphins are, in comparison, far more sociable and extrovert. They live in larger pods, and often breach, leaping partly or wholly out of the water, and riding the bow wave of passing boats and ships. As their name suggests, they have a distinctively shaped snout, though this is not always easy to see at a distance. The largest concentrations of bottlenose dolphins found off our coasts are in the Moray Firth on the east coast of Scotland, and also in the south and west, off Cornwall and west Wales.

Seeing whales off our coasts is trickier. The most likely species to encounter is the minke whale, one of the smaller species (though still about 7m (23 ft) long and weighing more than 10 tonnes). Minke whales are a regular sight in the Minch (between the Inner and Outer Hebrides), where they often feed alongside large flocks of seabirds; a good way to locate them is to watch for a concentration of birds on or above the surface, and then look out for the tell-tale sign of the minke whale's curved fin.

Other species of whale and dolphin are mostly scarce and irregular visitors to British waters. Sometimes, though, they cause a surprise, as with the northern bottlenose whale, a deep ocean species, that got lost and found itself swimming up the River Thames a few years ago. Sadly, despite efforts to rescue it and return it to the open sea, it eventually died. Other scarce visitors include the beaked whales and, on very rare occasions, two wanderers from the Arctic, the beluga and narwhal.

Probably the most sought-after cetacean regularly occurring in our waters is the killer whale, or orca. Once seen, they are unmistakable: no other creature has that combination of large size and black and white pattern. Recently it has been discovered that pods of hundreds of killer whales are following the mackerel fishing fleet off the northern coasts of Scotland. They also occasionally venture closer inshore, and are regularly seen around Shetland, Orkney and the Western Isles.

Another cetacean hotspot is the Irish Sea, between west Wales and eastern Ireland. Recently large numbers of harbour porpoises and fin whales – the world's second-largest whale – have been found there, often allowing close approach.

To have the best chance of seeing these creatures in British waters, it is worth taking a special whale and dolphin cruise. These run through the spring and summer months from various ports, mainly in southwest England, Wales and western Scotland, where cetaceans are most likely to be found.

Minke whales are one of several species of whale regularly seen in British waters.

Bottlenose dolphins are a regular sight around our coasts, especially in Scotland's Moray Firth.

Insects
& other
Invertebrates

Butterflies

Britain may not have very many different kinds of butterflies – fewer than 60 resident species, compared with several hundred found in continental Europe – but they certainly punch above their weight, as butterflies are some of our best-known and best-loved wild creatures. Even people who have a phobia about most kinds of insects love butterflies, not only for their dazzling colours and attractive appearance, but also because they seem to symbolise the magic of sunny summer days. Sadly, the recent run of cloudy, damp summers has led to a decline in several of our commonest species, such as the small tortoiseshell. On the other hand, fine, early springs have benefited others, including the orange-tip, one of the earliest butterflies to emerge.

Like other creatures, butterflies have had to adapt their lifecycle to the earlier arrival of spring, so that species that once emerged in April or May are sometimes seen as early as March or even February. For some scarce species, such as the Duke of Burgundy, earlier emergence is allowing them to produce a second wave of adult butterflies later in the summer, helping to boost their fragile populations.

Butterflies such as the orange-tip (above) and painted lady (below) are often found feeding on nectar-bearing flowers in our gardens.

Butterflies have also experienced mixed fortunes as a result of loss of habitat and climate change. Specialised species such as the wood white, heath and marsh fritillaries and the Adonis and chalkhill blues are less able to adapt to rapid change, and in many cases are in decline, though careful management of woodland and chalk downland habitats may now be stemming the tide.

Numbers of high brown and pearl-bordered fritillaries have fallen so sharply that they have disappeared from many former haunts and now face extinction in Britain. And our only alpine butterfly, the mountain ringlet, may also eventually disappear here as its high-altitude habitat warms up and becomes unsuitable for this hardy little creature.

Meanwhile, one of our most widespread butterflies, the small tortoiseshell, has also suffered because of the arrival from the European mainland of a tiny parasitic fly that kills off the caterpillars before they can pupate and change into adults.

But it is not all bad news for Britain's butterflies. Common, adaptable species such as the comma and peacock are taking advantage of plentiful nectar in our gardens, and a general rise in air temperatures, to extend their ranges farther north, while longer, finer autumns are allowing red admirals to survive much longer than before.

Red admirals are even sometimes overwintering as adults, joining the comma, peacock, brimstone and small tortoiseshell in hideaways in our garden sheds and garages. Here they spend the winter with their wings folded, to fool predators into assuming that they are simply dead leaves.

Butterflies astonish us with their amazing lifecycle. No other living creature undergoes quite such a dramatic change as the caterpillar when it metamorphoses, first into a pupa and then into the winged adult.

They can also amaze us with their feats of endurance, as in summer 2009, when we experienced an invasion by tens of millions of painted ladies, which had flown all the way from southern Spain or North Africa to breed here.

Three of our best-known garden butterflies, the red admiral (above), small tortoiseshell (left) and peacock (below), are undergoing mixed fortunes as a result of a run of wet summers.

Many people who witnessed this mass arrival were unaware that butterflies migrate at all, let alone such vast distances. Some autumns even see the arrival of monarch butterflies – one of the most incredible global travellers – in southwest Britain, having been caught up in storms and blown across the Atlantic.

Some rare butterflies are also crossing the Channel more regularly than before. The large tortoiseshell (once a native British species) and the Queen of Spain fritillary (a new colonist) have both been seen in southern counties in the past few years, and may soon become a permanent fixture here.

So, the future for Britain's butterflies is set to be a mixed one. We may see some species disappear, while others colonise, but one thing's for sure: they will continue to fascinate us with their lifecycle, and delight us with their beauty.

Moths

Britain may have only a few dozen kinds of butterfly compared with about 2,500 species of moth, yet moths are ignored, feared and even loathed by many people. This is partly because of their mysterious nocturnal habits, but also because the larvae of a handful of species, the dreaded clothes moths, make holes in our best woollen jumpers.

In fact not only do the vast majority of moths never feed on the contents of our wardrobes, but many species are not even nocturnal. These include the burnet and tiger moths, both of which can be found in areas of grassland and downland on warm summer days, and which are so active and colourful they are often mistaken for butterflies.

Several moth species such as the cream-spot tiger (above) and five-spot burnet (below) generally fly by day rather than after dark.

The difference between most moths and butterflies is that the antennae of moths are feathered, rather than clubbed (with small round knobs on the end), though there are some exceptions to this rule. Moths also usually rest with their wings folded or laid flat to the side of their body, whereas butterflies often close their wings above their body, but again there are many exceptions. So, given that moths and butterflies are so similar in many ways, perhaps we should learn to love moths a little more.

They certainly have fascinating lifecycles. Like butterflies, moths have four stages of development: egg, caterpillar, pupa and flying adult, with the adult stage sometimes being the shortest, lasting only a couple of weeks. So short are the lives of many adult moths that they do not even feed at this stage of their lifecycle.

There are many different groups of moths, from the tiny (and often almost impossible to identify) little ones, known as micros, through a wonderful array of names – pugs, carpets, thorns, sallows, brindles and beauties – to the huge and impressive hawkmoths.

This family includes some of our largest and most colourful insects, such as the pink and lime-green elephant hawkmoth (so named because the caterpillar waves its 'trunk' like an elephant to ward off predators), the lime and privet hawkmoths, and the legendary death's head hawkmoth, a scarce immigrant from the south that really does have a pattern resembling a human skull on the back of its head. This extraordinary insect frightens would-be predators by making a high-pitched squealing sound when it is disturbed, a sound that, together with its bizarre pattern, has led to many myths about the creature.

Another large and distinctive species, the hummingbird hawkmoth, is often responsible for mistaken claims of hummingbirds feeding on nectar in English gardens. In fact this creature, which does bear a remarkable resemblance to a hummingbird in shape, colour and habits, is a scarce migrant from the south of Europe and North Africa. In recent years, it has become a far more regular sight in southern Britain, as it takes advantage of climate change to extend its range northwards.

Moths are well known to be attracted to light, and this can help us get a closer look at some of our more mysterious, nocturnal species. If you don't have access to a proper moth trap, you can make an excellent substitute by shining a bright torch onto a white bed sheet. Moths are drawn to bright light because they use the moon to navigate, so by turning on the torch, you begin to draw them in. Once they reach the white sheet, they assume it is daytime, and look for a place to rest. Then you can easily catch them with a butterfly net and take a long, close look before releasing them.

A good moth identification guide is essential and will help you work out what you have caught. Many moths have extraordinary names – double kidney, Chinese character, ghost moth, drinker, steamer, spinach, burnished brass, spruce carpet, bleached pug, Kentish glory, swallow prominent, gothic, Blair's shoulder-knot, Brighton wainscot, exile, scarce merveille du jour, and setaceous Hebrew character are just a few!

The yellow-tail (left) has a rather odd yellow protuberance, which sticks out from between its wings, making it easy to identify.

Few moths are as large or impressive as the hawkmoths, including the eyed hawkmoth (above) and hummingbird hawkmoth (below).

Dragonflies & Damselflies

Dragons and damsels, as they are often known, are some of our most fascinating and impressive insects. Few others have quite such an extraordinary turn of speed and aerial manoeuvrability as the larger dragonflies, or the grace and beauty of some of the smaller, and often-overlooked, damselflies.

The difference between the two groups is easy to spot. When perched, all dragonflies hold their wings out at a 90-degree angle to their body, a bit like a World War I biplane (except the pairs of wings are one behind the other, rather than at the top and bottom). Almost all damselflies, on the other hand, hold their wings folded tight along the length of their bodies when perched, giving them a matchstick-like appearance and leading to the folk-name 'devil's darning-needle'. The only exceptions are the emerald damselflies, which perch with their wings held at a 45-degree angle to their body.

Damselflies are also mostly smaller than dragonflies, though the two species of demoiselle – beautiful and banded – rival the smaller dragonflies in size. These two species are generally seen in midsummer as they flutter across the surface of rivers and streams, their bright, iridescent blue-green shades giving them the folk-name of 'water butterflies'.

The emperor is Britain's largest species of dragonfly.

Other ways of identifying different species of dragonfly are to count the number of spots on their wings (though, confusingly, the four-spotted chaser actually has eight spots!), and to look closely at the shape of their bodies. Some species such as the hawkers are long and slender, while others, notably the chasers, are broad and squat. The colour of their bodies and wings is also an important way to tell them apart.

Our largest species of dragonfly, the emperor and golden-ringed, are among our largest insects. They are also the fastest flying, reaching speeds of up to 48 kph (30 mph), and able to turn on a sixpence in midair when chasing and catching their prey. Although they have large jaws and a long, pointed abdomen, they, like damselflies, are unable to bite humans, and do not carry a sting in the 'tail'.

Like butterflies, different species can be seen at different times of the year: the hairy dragonfly and some

of the damselflies are the first to emerge, on fine days in April; most damselflies then appear from May to July; the various darter and hawker dragonflies are mainly seen later in the summer, from July through to October, or even November in mild autumns. Their habit of basking in the sun – sometimes on the stems of reeds, but often simply on a path or boardwalk – makes them ideal subjects for photography, especially on chilly autumn days when they need to warm up before they can hunt for food.

Like so many insects, dragonflies and damselflies spend most of their lives in the early stages of their lifecycle. Eggs are laid on aquatic vegetation or deposited in the water, and the larva, or 'nymph', spends up to seven years under the surface, preying on smaller aquatic insects and other invertebrates, which it kills and despatches with its fierce and powerful jaws. Finally, on a fine, warm day in spring or summer, the adult dragonfly or damselfly emerges from the water, usually crawling up the stem of an aquatic plant, before drying its wings and flying away.

The four-spotted chaser is one of our commonest summer dragonflies – but note that it actually has eight spots on the wings, not four!

Adult dragonflies are also very effective predators. Their fearsome jaws enable them to catch a wide range of smaller insects, including damselflies. However, in turn, they provide food for others, including the dashing hobby, a falcon that catches dragonflies in midair with its sharp talons.

Other threats faced by these insects are water pollution and habitat loss through the drainage of ponds and lakes, many of which have been lost in the past few decades. Fortunately, though, the project to restore many wildlife-rich habitats – especially watery ones – is giving dragons and damsels a helping hand.

Climate change is also enabling these very mobile and adaptable insects to push farther north in Britain, and has allowed new species such as the willow emerald to colonise southeast England in the past decade. The next few years look set to bring even more species across the Channel, to the delight of fans of these extraordinary creatures.

The banded demoiselle is one of the largest and most attractive of our damselflies.

Crickets & Grasshoppers

Crickets and grasshoppers are often confused with one another. They are both small, brownish-green insects that hide in long grass, and when they are spotted, they jump away on spring-loaded legs. But there are important differences between the two groups.

First, their appearance. The antennae of grasshoppers – the important sensory organs with which they detect movement, sound and smell – are generally much shorter than those of crickets, which can be even longer than their body. Likewise, the ovipositor of female grasshoppers – the long tube at the end of the abdomen with which she lays her eggs – is also shorter than that of a cricket.

Their habits are different, too. Most grasshoppers are diurnal, active during the hours of daylight, whereas crickets tend to be more nocturnal or crepuscular (active around dawn and dusk). They also feed on different things. Grasshoppers are herbivorous, generally eating grass, while crickets are far more omnivorous and predatory, hunting down smaller insects, which may include grasshoppers, and swiftly despatching them with those fearsome jaws.

But the biggest difference between these two closely related groups of insects is in the way they sing – those amazingly loud and far-carrying sounds, known as stridulations. Grasshoppers stridulate by rubbing their rear legs against their forewings, in a rather contorted action. Crickets, on the other hand, do so by rubbing their forewings together. Both produce extraordinary sounds, designed, like birdsong, to repel rival males and attract a female.

There are eleven species of grasshopper in Britain, four species of true cricket and eleven bush crickets. The latter include some of our largest and most impressive insects: the great green bush cricket and the wart-biter cricket. Found in grassy areas throughout southern Britain, the great green bush cricket can grow up to 45 mm long (1¾ in). Together with its bright green colour and noisy chirping, this makes it a reasonably easy species to find.

The same cannot be said of its close relative the wart-biter cricket, a chalk grassland specialist that can now be found on a few scattered downland sites in southern Britain, from Dorset and Wiltshire in the west to Sussex and Kent in the east. This large insect was so-named because it was once thought it could cure warts by biting them off – presumably a fairly painful remedy.

All four of the true crickets are relatively scarce, including the house cricket, which was introduced here many centuries ago. The field cricket is one of Britain's rarest insects, found only on a few sites in Surrey, Sussex and Hampshire. It is currently being given a helping hand by being bred in captivity and released at selected sites, including the cricket ground at Arundel.

Grasshoppers are generally far more common and widespread than crickets, and are found throughout most of lowland Britain. They tend to be fairly small – about 12–20 mm (½–¾ in) long, with females usually larger than the males – and a combination of dull brown, grey and pale green in colour. Common species include the meadow, field, mottled and common green grasshoppers. These are found in most places, whereas others, including the large marsh, lesser-mottled and stripe-winged, are far more localised.

Both crickets and grasshoppers belong to the group Orthoptera; other relatives include cockroaches and earwigs. These often repel people, but we should also admire them: cockroaches for their legendary survival skills (it is often said that they would survive a nuclear war) and earwigs for their maternal instincts. The mother earwig will not only guard her precious eggs from any hungry predator but she will also clean them regularly to stop mould forming on their surface – a wise precaution in the damp, cool places where earwigs usually live.

The meadow grasshopper is one of our commonest and most widespread members of its family, found throughout Britain but not in Ireland.

Beetles

When asked what his long years of nature study had taught him about the existence of God, the great biologist J.B.S. Haldane paused for a moment and then replied: 'That he had an inordinate fondness for beetles!'

Haldane's witty reply had a serious side, for of all the known animal species on Earth, approximately one in four, or 25 per cent, are beetles. When it comes to insects, beetles are even more dominant: around 40 per cent of all known insect species are beetles, with approximately 400,000 species in all. Yet compared to the more showy butterflies or moths, or dazzling dragonflies and damselflies, we often overlook them.

Beetles come in all shapes and sizes, from the small and colourful ladybirds (above and right) to the fearsome stag beetles (below).

The success of beetles is hardly surprising given their adaptability and toughness. The 4,000 or so different kinds of beetle in Britain can be found in almost all of our habitats, apart from the high tops of the Scottish Highlands or in our marine environments. They are particularly adapted to life in our woods and forests, living in the leaf litter or tree canopy, where they feed on either fungi or plants, or hunt down other insects and invertebrates. A few live underwater, such as the great diving beetle, one of our largest insects. This aquatic monster preys on other freshwater creatures, and has the ability to form an air bubble around its body to enable it to spend long periods underwater.

Other members of this diverse group of insects include cockchafers and rose chafers, ground and leaf beetles, sexton beetles (which bury a corpse of a small mammal into which they lay their eggs), and two notorious creatures that every householder dreads: woodworm and deathwatch beetle. The latter got its name from the strange knocking sound it makes at night, which was associated with the time when relatives would sit with the dying.

Of all our beetles, probably the best known and best loved are the ladybirds. Britain has 25 or so ladybird species, which vary in both their colour and pattern. They are loved by gardeners for their habit of preying on aphids, and so are usually

welcomed, though in some years, such as the long, hot summer of 1976, they can appear in plague-like proportions in some places.

In recent years, an alien invader, the notorious harlequin ladybird, has arrived here. This large and very variable ladybird – individuals can be any combination of black, red and orange and have a varying number of spots – entered the UK on plant material imported by nurseries and garden centres, and in less than a decade has spread across much of the country. It predates on a wide range of insects, including smaller species of ladybird, which means that it is now threatening several of our native species.

Our largest native beetle is also one of the scarcest: the stag beetle. This huge creature is mainly confined to the south and east of Britain, as it needs a warm climate, which means that one of its strongholds is the London suburbs, where it can often be seen flying on warm nights in June or July. Male stag beetles are magnificent specimens, with huge claws with which they will fight their rivals to gain the affection of the watching female – much as rutting deer stags do, only in summer rather than autumn. The stag beetle is also associated with a wide body of folklore, including the mistaken belief that male stag beetles would carry hot coals in their claws, then drop them onto thatched roofs, setting them alight.

The male glow-worm (above) lacks the female's ability to glow, but, unlike her, he can fly.

Another kind of beetle long associated with light is the glow-worm. The flightless female glow-worm emits a bright light from its rear end, which attracts passing males as they fly by. Glow-worms have declined in recent years because of loss of habitat and also light pollution, which makes it harder for the males to find their mates – and harder for us to find them too.

The golden-bloomed grey longhorn beetle (left) is one of our most attractive beetles, and can be found in meadows and hedgerows, mainly in southern Britain.

Bees & Wasps

Bees and, particularly, wasps may not be the most popular insects but they are certainly among the most useful. It is said – with some justification – that if honeybees and bumblebees stopped pollinating our fruit, vegetables and other crops, human civilisation would soon grind to a halt.

For centuries, we have taken advantage of honeybees' sociable habits and ability to make honey by domesticating them and keeping them in hives. However, in recent years, problems with parasites such as the varroa mite, and a mysterious phenomenon known as colony collapse disorder, have led to many hives being deserted.

There are almost 300 different species of bee in Britain. These fall into three main categories: honeybees, bumblebees and solitary bees, which are by far the largest group, with about 250 different species, including mason bees, leafcutter bees and carder bees. Solitary bees, as their name suggests, do not live in the large, social colonies favoured by honeybees. Instead, they build a single-celled nest where they lay their eggs, often in a crevice in a wall or trunk of a tree, or in another part of a wood or garden.

There are about 20 species of bumblebee in Britain, of which half a dozen are common and widespread. They, too, are solitary. These include *Bombus terrestris*, also known as the buff-tailed bumblebee, which is one of the earliest species to emerge in spring, and is often seen on a fine and sunny day feeding on nectar in gardens. In recent years, especially during mild winters, they may emerge even in the middle of winter, feed for a while, then return to their hiding places when the weather turns cold again.

Other common and familiar bumblebees include the white-tailed and red-tailed bumblebees, whose names help you identify them. The red-tailed also has a black, fluffy body, which looks a bit like a guardsman's busby. Some species of bumblebee lay their eggs in other bees' nests, allowing them to do all the hard work in raising their offspring. Not surprisingly, they are known as cuckoo bees. If you look closely, you can see that

Honeybees, wasps and bumblebees may sometimes scare people but they do a vital job as pollinators.

they do not carry pollen on their hind legs.

Wasps are far less popular than bumblebees, yet they, too, provide a useful service, predating on smaller insects such as flies, which otherwise might become too numerous and spread disease. Next time wasps disturb your summer picnic, think of them as nature's pest controllers, and you may feel better about them. In fact wasps rarely sting unless provoked, though beware of disturbing them at their nest, as they can turn very nasty. Unlike honeybees, which die once they have used their sting, wasps are able to sting more than once, as can bumblebees.

Like bees, wasps may be social or solitary. Social wasps build huge nests out of wood pulp, effectively papier-mâché, while solitary wasps usually do not build nests at all. Exceptions include the potter wasp, which builds an extraordinary pot-shaped nest out of sand and clay. Not all solitary wasps have the classic 'yellow-and-black' appearance, and many other insects, notably hoverflies, mimic wasps in order to avoid being eaten by predators, so to identify them you need to look closely.

Many wasp species are parasitic, laying their eggs inside the eggs of other insects. Their larvae live off the host, eating it away from within until it dies, then change into the adult and fly off to begin the cycle all over again.

The largest of the UK's wasp species is the hornet, which, at over 3 cm (more than 1 in) long, is one of our most impressive insects. Despite their size, hornets are placid creatures, which rarely attack humans, though their sting is a powerful one. In the past few years, hornets have suffered declines in their range and population – they need dead or decaying trees for nesting places but modern forestry practices tend to remove these for safety reasons.

The hornet clearwing is one of the finest examples of mimicry in nature – looking remarkably like the hornet; this is in fact a moth.

The buff-tailed bumblebee is one of the commonest of more than 20 species in its family, many of which are threatened by loss of habitat.

Ants

Ants are one of the most highly developed groups of social insects on the planet, with a well-deserved reputation for cooperation, organisation and sheer strength – an ant can carry well over 100 times its own weight. They are related to wasps and bees, in the order *Hymenoptera*.

There are at least 12,000 known species of ant in the world, and probably many more are yet to be discovered, but in the UK we only have about 50 different kinds. Of these, by far the most common and familiar is the black garden ant. These have adapted very well to living in gardens, often nesting in cracks between paving stones, and frequently entering our homes in order to find food. As a result, many people regard them as a pest, and fail to recognise that they are one of the most complex and fascinating of all our native creatures.

Black garden ants live in colonies of several thousand individuals, mainly 'worker ants', which are infertile, wingless females. As their name suggests, they do most of the work required to keep the colony going. There are also males (drones) and one or more fertile females, known as queens. Black ants feed both on nectar and on other insects, dragging their prey down to their colony in a team effort. On warm summer days, look out for flying ants – mostly males – that emerge to search for and mate with the queens.

Other familiar garden ants are the yellow meadow ant and the red ant. The former prefers wilder areas with longer grass, where it builds its large, mound-like nests, which may be mistaken for molehills, though they are usually covered with grass. These may attract one of the ants' main predators, the green woodpecker, which uses its long, sticky tongue to vacuum up ants by the hundred.

Yellow meadow ants spend most of the time underground, feeding on any insects they find in their network of tunnels. In contrast to the black garden and yellow meadow ants, which do not hurt humans, the various species of red ant have a painful sting. They live in smaller colonies of a few hundred individuals, sometimes including many queens.

Of all our ants, perhaps the most fascinating – and certainly the most impressive nest-builders – are the wood ants. Three species occur commonly in Britain; the southern wood ant, found mostly in southern England and Wales; and the northern and Scottish wood ants, found, as their name suggests, farther north. Wood ants build extraordinary dome-shaped nests, up to a metre (3 ft) across, on the forest floor.

At first sight, these nests look like a heap of leaves, twigs and (in coniferous forests) pine needles, but on a warm spring or summer's day, if you look more closely, you may see large numbers of ants crawling rapidly across the surface of the dome. They leave the nest to feed on honeydew from aphids on nearby trees, and will also eat other small insects that come their way.

Wood ants' nests are usually in clearings or on forest rides and paths, where the sunlight can warm them up – these ants are an indication that the wood is well managed, as neglected woods soon lose these sunlit patches. The ants can regulate the amount of warmth entering the chambers inside the nest by opening or closing the entrances to their tunnels – a kind of natural air-conditioning system.

If you get too close to a wood ants' nest, watch out! The ants may assume you are an intruder, and squirt formic acid at you – painful if it gets in your eye or onto your skin. One experiment to try is to put a piece of litmus paper near the nest's surface to provoke the ants. As the formic acid hits the surface of the paper, it will turn red to indicate the presence of an acidic substance.

This black garden ant is 'herding' aphids, before milking them for their honeydew.

Spiders

Unlike bees, wasps and ants, spiders are not insects, but arachnids, related to scorpions, mites and ticks. Unlike insects, they have eight legs in four pairs, not six, and a two-part, rather than three-part, body. Instead of antennae, they have a series of bristly hairs along their legs and body, which can pick up vibrations, smells and even flavours – all very useful when hunting down their prey.

All spiders are carnivorous in their diet, using their fangs to inject poison into their victims' bodies before eating them. Some catch their prey using complex webs. Spun from silk, which is so strong it can bear more weight than its equivalent thickness in steel, the webs are covered with sticky blobs that trap any creature unfortunate enough to land on them. The spider sits in the centre, and uses its sensitivity to movement to notice the vibrations when prey lands in the web, enabling it to reach it quickly and put it out of its misery. Others are ambush hunters, while some are stalkers. All have the ability to scare some people, whose fear of spiders has the scientific name 'arachnophobia'.

There are over 600 different species of spider in Britain. Some are so small they are hardly visible to the naked eye, while others have a body length of several centimetres, plus equally long legs, making them appear rather fearsome. The house spiders of the genus *Tegenaria* live all year round in our homes but are especially noticeable in early autumn, when males go on the move in search of females to mate with. At this time of year, they may be glimpsed running rapidly across a carpet or found stuck in the bath – their legs are unable to grip to climb up the smooth sides, and they need to be rescued before they drown.

Garden spiders spin intricate webs to trap their insect prey.

One unusual species also found in our homes is the daddy-long-legs spider, confusingly named after a familiar insect, the daddy-long-legs, or crane fly. Daddy-long-legs spiders are so slender that it is easy to assume that their rather untidy web is empty. Or at least it appears so until you tap it with your finger, at which point the spider vibrates rapidly backwards and forwards faster than your eye can see, a clever ruse to avoid being caught by a predator. Daddy-long-legs spiders probably came to Britain with human visitors from continental Europe, and have spread northwards by the simple but effective means of hitching a lift inside furniture when people move home!

The widespread theory that the daddy-long-legs spider has the most venomous bite of any British spider is in fact an urban myth. However, although the vast majority of spiders do not bite humans, more than a dozen species found in Britain, including some native ones, have been known to do so, sometimes inflicting quite a painful bite. Some exotic foreign spiders, including tarantulas, may be imported on fruit such as bananas but usually they do not survive very long in the British climate.

Outside the home, the most familiar spider is the garden spider, which builds a classic orb-shaped web, easily visible when covered with dew on autumn mornings. Another autumnal spectacle happens when thousands of tiny money spiders spin their floating strands of web, known as gossamer. These allow the spiders and their offspring to float through the air for long distances, seeking out new homes. If you are lucky enough to witness this, it is most impressive.

Some of the most impressive British spiders are the wolf spiders, which perform the most extraordinary courtship display, the male waving his extended, arm-like mouthparts in a complex series of semaphore-like moves in order to impress the watching female.

Other unusual spider behaviour includes living in water. The great raft, or fen raft, spider sits half-submerged on the surface and hunts by descending into the water or grabbing passing prey. Found in only three locations in Britain – Suffolk, East Sussex and South Wales – it is one of our largest spiders, with a body measuring 2 cm (¾ in) long and with an overall span (including its legs) of up to 7 cm (2¾ in).

The raft spider lives a semi-aquatic lifestyle in the waterways of East Anglia.

Woodlice, Slugs & Snails

Unlike slugs, snails carry around their home with them in the shape of their protective shell.

Woodlice are curious creatures, neither bugs nor beetles, which they superficially resemble, but crustaceans. Their relatives include crabs, lobsters and shrimps. Unlike most of their relatives, however, they have adapted to life on land rather than in fresh or salt water, though they retain many characteristics of aquatic animals, including a need to live in cool, damp places – expose a woodlouse to direct sunlight and it will soon dry out and die.

About 40 species of woodlouse are found in Britain. They are familiar creatures, and easy to find, usually beneath stones and rocks. Woodlice are eaten by many different creatures, though only one species, the appropriately named woodlouse spider, preys exclusively on them, piercing through the woodlouse's armoured shell with its prominent fangs to get at the soft interior.

After she lays her eggs, the female woodlouse will keep them attached to the underside of her body in a kind of pouch, out of which her offspring eventually hatch. Woodlice feed mainly on dead plant material, making them a crucial participant in healthy ecosystems, as they recycle nutrients into the earth.

Perhaps because they are so common in our gardens and are completely harmless to humans, woodlice have gained a special place in our affections. We give them familiar names, which either relate to their ability to roll up tight into a ball as a defence against predators or, rather strangely, compare them to pigs. Names include pill bug, armadillo bug, roll-up bug, sow bug and chuggypig.

In ancient times, woodlice were used as pills and given to patients with stomach ailments or heartburn. There is some scientific basis to this, as the woodlouse's skeleton is mainly made up of calcium carbonate, which can neutralise the stomach's acids.

Snails and slugs are less popular with householders and gardeners, mainly because they feed on living plants, and so can cause extensive and repeated damage to prized specimens. Both are molluscs, but with very different lifestyles: the snail carries his home with him, and can retreat inside his shell at the first sign of danger, while slugs are defenceless against predators.

Slugs vary dramatically in size; some are among our largest molluscs, reaching lengths of 15 cm (6 in) or more. One of the most impressive is the leopard slug, which really does have a pattern and coloration rather like a leopard. Other giants include the large black slug, a common garden resident. All these slugs are most active on warm, damp nights in summer; the next morning, you may find tell-tale trails of sticky mucus on your garden paving or even inside your home – slugs are able to squeeze through the tiniest gaps.

Because of their slimy appearance, slugs may not be very popular, but they do provide valuable food for many of our favourite garden creatures, including the song thrush and hedgehog. Both these declining species act as natural pest controllers and remove the need to use slug pellets to keep these molluscs under control.

The most frequent snail species usually encountered is the garden snail, which, despite its reputation for slowness, can travel long distances under cover of darkness. A much rarer creature, mainly found in southern England, is the introduced edible, or Roman, snail. As both names suggest, the Romans, who prized them as a delicacy, originally brought these huge molluscs to Britain. Another common sight, especially in grassy areas of the countryside, is white-lipped snails. These have an amazingly varied appearance – specimens with very different colours and patterns are often found alongside each other – but all belong to the same species.

Some of our most fascinating snails are found underwater. These include the cherrystone snail, great ramshorn snail and the great pond snail, which can often be seen feeding upside down on the surface of ponds or lakes.

Despite their terrestrial existence woodlice are crustaceans, closely related to lobsters and crabs.

Rockpool Life

Starfish (right) and edible crabs (below) are two of the commonest inhabitants of seaside rockpools.

No visit to the British seaside is complete without a spot of rockpooling – exploring the temporary pools that provide a welcome oasis for marine creatures when the tide goes out. Unlike sunbathing, the good news is that you can do it all year round.

The life in rockpools is almost as varied as that in the sea itself, albeit in a more miniature form. A wide range of fish, molluscs and crustaceans, of all shapes and sizes, can be found, creating a fascinating ecosystem where you can enjoy really close-up views.

The commonest fish in rockpools are gobies, blennies, bullheads, lumpsuckers and rocklings. They often live in these pools permanently, hiding in the sand or in a rocky crevice, where they feed on smaller fish or other marine creatures such as shrimps. Other fish found here usually live in the more open areas of sea but become trapped at low tide. Common blennies are generally greenish-brown in colour, while rock gobies are paler brown and often show dark bands along their body. Bullheads, also known as 'miller's thumbs', have a characteristically large head, while lumpsuckers are round fish with a round sucker on their underside, which they use to attach themselves to the bottom of the rockpool. Rocklings are longer and more slender in shape, with a large mouth and mottling along the body.

Crustaceans such as crabs are among the most obvious rockpool residents. They include the common shore crab, which is quite variable in colour, ranging from greyish-black to a paler olive-green, and the edible crab, which can grow to quite a large size and is easily identified by the crimped edge to its shell, rather like that of a Cornish pasty! Also look out for the hermit crab. This fascinating creature chooses to live inside the discarded shells of other marine animals such as whelks, changing

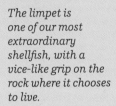

its home as it grows to obtain more roomy accommodation. Hermit crabs have one claw larger than the other, and use this to block the entrance to their shell, to prevent a predator from getting inside and scooping them out.

Other crustaceans commonly found in rockpools include shrimps, which scavenge for tiny morsels of food. Living shrimps are brown rather than the familiar pink colour of the cooked version, and can be surprisingly hard to see as they hide away in the sand. Look for tiny movements as they wriggle downwards.

Some of the most fascinating rockpool creatures don't actually look like animals at all. As their name suggests, sea anemones are rather like exotic flowers, which come in a bewildering range of colours. They are related to jellyfish, and share many of that group's characteristics, including a soft body. However, be careful not to touch them – they can give a nasty sting. Starfish, another soft-bodied group of creatures, are scavengers as well as predators of their fellow rockpool inhabitants. They typically have five arms, but vary considerably in shape and colour.

The limpet is one of our most extraordinary shellfish, with a vice-like grip on the rock where it chooses to live.

Other soft-bodied creatures protect themselves with a hard shell. These include cockles, mussels and whelks, and those that hold fast to a rock – limpets and the smaller barnacles. Don't be fooled, though, by the static limpets you see at low tide. When the sea comes in, they float off their resting places and move around the rock surface, vacuuming up tiny pieces of food, before returning to exactly the same place as they started.

Seaweed is another fascinating feature of rockpools that should not be overlooked. It comes in a wide range of different species, including every child's favourite, the bladderwrack, whose floating air bladders can be popped with a satisfying noise.

Unlike other crabs, the hermit crab lives in the discarded shells of other marine creatures.

Exploring rockpools – with or without children – is great fun, and you just need a fishing net, a bucket or tray to keep your specimens in, and a magnifying glass for taking a close look at the smaller creatures you find. Polarising sunglasses can also be useful on a sunny day, as they let you look beneath the surface without being blinded by the sun's glare.

When rockpooling, always keep a close eye on the sea. When captivated by these incredible creatures, it is easy to overlook the reason why they are there in the first place – the twice-daily movement of the tides.

Reptiles, Amphibians & Fish

Snakes

Compared with continental Europe, Britain has a rather poor number of 'herps', as the experts call snakes and lizards, with just three species of snake and three native lizards.

The reason for this is that as Europe's fauna began to head back northwards after the end of the last Ice Age, many species did not quite make it to Britain before the English Channel created a barrier between the continent and us. The situation in Ireland is even worse. As followers of the legend of St Patrick know, there are no snakes in Ireland because the saint supposedly drove them all out. In fact the Irish Sea flooded the land between Britain and Ireland even earlier, well before any snakes had the chance to get there.

The three species of snake found in Britain itself are the grass snake, adder and smooth snake. All are small to medium-sized, mostly about 50–75 cm (20–30 in) long, though the grass snake can grow to a length of 1.5 m (5 ft).

Of our three native snakes, the grass snake is by far the most common and widespread, being found throughout most of lowland England and Wales, especially in areas near water, though it is extremely rare in Scotland. It is a greenish-brown colour, with dark marks along its sides, and a distinctive yellow band across the collar. The most aquatic of our snakes, it is an excellent swimmer, and may be seen with its head sticking out of the water as it swims along in search of frogs and toads, its main prey.

Grass snakes lay up to 40 eggs in the warmest place they can find, usually a heap of rotting vegetation by the side of a river, pond or lake, but also frequently in man-made compost heaps. The young hatch ten weeks later and are immediately able to fend for themselves.

The adder is smaller and chunkier in shape than the grass snake, and is our only venomous reptile. However, deaths from their bites are extremely rare, and the snakes themselves are keen to avoid humans and will bite only under extreme provocation. They can be easily identified by the distinctive zigzag pattern down their brownish back, which can stand out beautifully against a paler background. Dry-country reptiles, adders are found on heathland and moorland, and on the edges of woodland,

Grass snakes are the most aquatic of all our reptiles, and often hunt while swimming.

throughout Britain. They emerge very early in the spring, on the first warm, sunny day of the year, and will flatten themselves to soak up the sun's weak rays in order to warm up. The females breed every two or three years, the young being born covered with a thin membrane, which ruptures at birth.

Adders prey mainly on small mammals such as voles and mice, and will also take the chicks of ground-nesting birds. Their venom paralyses the victim so it can be subdued and swallowed with ease.

The third British species, the smooth snake, is also by far the rarest, found only on the dry, sandy heaths of Hampshire, Dorset and southwest Surrey, though the species is currently being reintroduced into Devon. It is quite similar in appearance to the adder, but more slender and without the distinctive zigzag pattern on its back, and with a round pupil (rather than the adder's vertical one). Smooth snakes prey on other reptiles, including the sand lizard and slow-worm, which share its heathland habitat.

All British snakes shed their skin to enable them to grow, and cast-off skins may sometimes be found. They also all hibernate, from October through to March or April.

The best time of year to look for snakes is in spring, just after they have emerged from hibernation, or autumn, just before they begin to hibernate again. Then, the temperatures are much lower than in summer. Early in the morning, when the snakes often show themselves because they need to warm up, the cold makes them much slower, giving you a chance to get prolonged views. Another way to find snakes is to put pieces of corrugated iron in suitable habitats, wait for a few days, then lift them. Sometimes you will be rewarded by the presence of a snake, though you will need to be quick before they slither away to find another hiding place.

The adder (above) and smooth snake (below) are commonly found on heaths in southern Britain, such as the New Forest.

Lizards

There are three species of lizard native to Britain – the common and sand lizards and the slow-worm – but there are also at least two other species – the wall and green lizards – found here in small numbers.

Of the three native species, the commonest and most widespread is the slow-worm. Despite its appearance, this is not a snake but a legless lizard, which evolved from ancestors with legs but lost them over time.

Much smaller than any of our native snakes, slow-worms are usually about 20 cm (8 in) long, and about the width of a human finger. They can also be told apart from snakes by their ability to blink, which snakes cannot do because they have no eyelids. Like other lizards, the slow-worm can jettison its tail if it thinks it is in mortal danger, confusing its predator. Indeed, its scientific name, *Anguis fragilis*, means 'fragile snake'.

Slow-worms are generally golden-brown in colour, with dark spotting along the length of their body, though in the sun they can appear rather like burnished bronze, while young slow-worms are a paler, golden shade. They prey on slugs and worms, making them a natural pest controller, so are welcomed by gardeners. The name slow-worm derives from its ability to kill its favourite food: it means 'slay-worm' and has nothing to do with the animal's speed!

Sand lizards often bask in the sunshine to warm up early in the day.

Although slow-worms are widespread in a range of habitats, they are not always very easy to see – having many predators means they tend to lie low. They can be found in many gardens, especially those with a compost heap, where they often nest, like grass snakes, giving birth to live young. They also hide away in log piles and rockeries.

Common lizards also give birth to live young. Indeed, their alternative name, viviparous lizard, means just that. They are found on mainly dry, lowland grassland, including heaths, moors and coastal cliffs, throughout much of Britain, and are also the only lizard to be found in Ireland. They are best looked for as they warm themselves up on grassy, south-facing banks,

on sunny days from March through to October, after which they hibernate for the winter.

Unlike our other native species of lizard, sand lizards are scarce and restricted in both range and habitat. They are confined to two regions of England: on the sandy heaths of Dorset and Hampshire, and the coastal dunes of Lancashire, where there is a smaller population. They have also been introduced to the island of Coll in the Inner Hebrides.

Common and sand lizards can be hard to tell apart. Both sexes are brown but the male common lizard can have vivid green markings, while its sand lizard counterpart is usually even more dazzling in appearance. It is also rather bulkier, with a larger head.

Two European species of lizard can also be found in a few places in southern Britain. The green lizard is a large and striking reptile, growing up to 35 cm (14 in) long, making it by far the largest lizard found here. Green lizards are common on the Channel Islands, where they are native, and can also be found in Dorset, on the cliffs at Bournemouth, where they were illegally introduced some years ago. This is one of the few places warm and sunny enough to support this species, but as global warming takes hold, it may be able to spread farther north.

The other introduced lizard species, the wall lizard, can be found at 40 sites in southern Britain. It is similar to the common lizard in size and coloration, though it can be told apart by its more pointed head, and longer body and tail. Given the success of a population of wall lizards introduced into North America in the 1950s – they have already spread some distance from their original release location – it is possible that the wall lizard could become a much more frequent sight in Britain in the next few decades.

The wall lizard can appear remarkably similar to the native common lizard – this is a female.

The slow-worm is Britain's commonest and most widespread species of lizard.

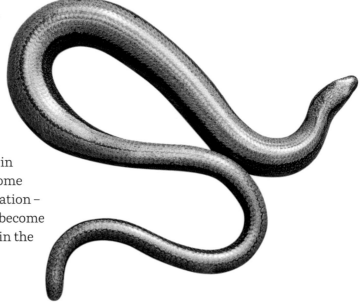

Newts

Newts are fascinating creatures. Although superficially lizard-like in appearance, with a slender body and long tail, they are not reptiles but amphibians, as they spend part of their lifecycle on land, and part in water. Like frogs and toads, they have a number of special adaptations to this semi-terrestrial, semi-aquatic way of life.

There are four different stages to the newt's lifecycle. They begin life as eggs, which are laid separately (unlike frogs' eggs, which are laid in clumps, and those of toads, which are laid in strings) and usually attached to aquatic plants underwater. The eggs hatch into larvae, known as tadpoles, which develop in their watery home until they have grown the legs they require for walking on land, and have replaced their gills (for breathing underwater) with lungs. During their development, they feed on small aquatic invertebrates and algae.

The third stage of the newt's lifecycle is the one we tend to overlook. This is when they leave the water and begin life as a terrestrial animal, known as an eft. Finally, they grow into the full-sized adult newt, which also lives mainly on land, only returning to the water each spring and early summer to breed and lay their eggs. The best time to see newts, which are well worth close study, is at this time, especially on warm days. During the winter months, they hibernate on land, often burrowing into log piles or rockeries, where they may be discovered in quite large numbers.

There are three species of newt native to the UK: the smooth (or common) newt, the palmate newt and the great crested newt. Both smooth and palmate newts are found in freshwater habitats throughout Britain, but the smooth newt is the only species found in Ireland.

Smooth and palmate newts are very similar in appearance, brownish in colour and about 5 or 6 cm (2 or 2½ in) long, although the former may grow as large as 10 cm (4 in). The males of both species develop a wavy crest along their back in the breeding season, when they can be confused with the great crested

During the breeding season the normally rather drab smooth newt dons his splendid breeding garb.

newt. The palmate newt has webbing on its back feet, which gives the species its name.

A male palmate newt, so named because of the webs between his toes.

The great crested newt is the largest and rarest British newt, and also the most impressive predator, able to feed on quite large prey, including water snails and tadpoles. They grow up to 15 cm (6 in) long, and look much bulkier than their smaller relatives. Dark in colour, they can appear almost black from above, and have a bright orange underside, spotted with black. In spring, the breeding season, the male sports a long, ragged crest all the way from its neck to the end of its tail.

Outside the breeding season, great crested newts, like other newts, live mainly on land. In recent years, the species has undergone a serious decline and is now carefully protected: it cannot be disturbed, taken into captivity or harmed in any way. As a result, land developers have a legal duty to look after great crested newts, which often entails costly schemes to relocate them to new habitats.

Newts of all species are excellent indicators of environmental change because of their complex habitat requirements. They are especially vulnerable to the acidification of their watery homes caused by air pollution; to pollution caused by the discharge of industrial or agricultural chemicals into rivers, ponds and streams; and to disease. In recent years, smooth newts in particular have found a refuge in our garden ponds. These usually provide a safe haven for them, though domestic cats can and do prey on them.

Frogs & Toads

Three species of frog and toad – the common frog, common toad and natterjack toad – are native to Britain. But it is also home to one introduced species, the marsh frog, and another that was almost certainly once native to Britain and is now the subject of a reintroduction programme, the pool frog.

Other non-native species, including the edible frog, European tree frog, and the fire-bellied, midwife and clawed toads, are also occasionally found here, thanks to deliberate or accidental releases into the wild. However, none of them has managed to spread significantly.

The common frog is, as the name suggests, our commonest and most widespread member of its family, being found in watery habitats throughout mainland Britain. In recent years, its population has suffered sharp declines, partly due to the draining of many farm ponds and other bodies of water in the wider countryside. During this period, the increase in the number of garden ponds has created a haven for this attractive amphibian, enabling it to survive in areas where water is otherwise scarce.

The natterjack toad (above) is easily identified by the narrow yellow stripe running down its spine.

The common frog (below) is a regular visitor to garden ponds, especially during the spawning season in late winter.

All Britain's amphibians are currently being threatened with chytridiomycosis, a disease caused by the chytrid fungus. This, together with pollution and climate change, may cause dramatic declines in our native populations during the coming decades.

The marsh frog was first brought to the UK in the 1930s, when some were released in Kent. Since then they have spread northwards and westwards and can now be seen – and heard, as they have a very loud and distinctive call – in ditches and dykes across much of southern England.

The pool frog has a chequered history as a British species. Almost certainly originally native, the last populations died out in East Anglia in the mid-1990s. There is now a scheme to reintroduce them at a secret site in Norfolk, from where it is hoped they will spread to other parts of the UK. They are smaller than the common frog but otherwise it is quite hard to distinguish them.

Given good views, it is easy to tell frogs apart from toads. Common frogs are usually brown, greenish or olive in colour, with varying amounts of dark blotches. They often show a dark mask, and have a smooth skin. Toads, in contrast, have a rough, rather warty skin, and are generally a more even brown in colour. They also walk rather than hop, and lay their eggs in strings rather than huge clumps, as frogs do.

The common toad is a more terrestrial animal than the common frog, and spends a large proportion of its life on land rather than in the water. Its skin appears drier than that of the frog, and it is more able to retain water so that it does not dehydrate. However, like all amphibians, toads prefer damp places.

Early in the breeding season – usually sometime in February – toads undertake the hazardous journey from their hibernating areas to their breeding grounds, which often entails crossing roads. In some locations where this is a regular annual occurrence, special toad crossings have been set up, to give the animals a helping hand and stop them being squashed by motor vehicles. Once the toads reach their breeding ground, the males will climb onto the back of a female and hang on for days on end, to prevent her from mating with a rival.

The other British toad, the natterjack, is far more rare and localised, being found in scattered sites across England and southwest Scotland. They are best identified by the conspicuous yellow stripe down the centre of their back, and also by the extraordinarily loud and rasping calls given by the males on warm spring evenings.

The marsh frog is a non-native species introduced from continental Europe, and found mainly in southern Britain.

The common toad has a drier, wartier skin than that of the common frog.

Freshwater Fish

The pike is our largest and most predatory freshwater fish.

Fishing for tiddlers in a local pond or stream was once a classic childhood activity. Although the tiddlers in question – usually minnows or sticklebacks – may be small, they have fascinating lifecycles, with sticklebacks, in particular, well worth a closer look.

In spring, at the start of the breeding season, the male stickleback turns a bright red colour underneath and defends his watery territory with a fierce determination. Once he has attracted passing females, and has fertilised their eggs, he then acts as sole parent, even making a little underwater nest in which he places the eggs, fanning them with his fins in order to keep them oxygenated. Once the 100 or so young have hatched, he continues his fatherly care until, after a week or so, they gain their independence and swim off on their own.

About 40 species of freshwater fish are native to our lakes, rivers and streams. These include several commonly sought-after by anglers, such as perch, roach, dace and chub, as well as the giant pike and more unusual creatures such as lamprey, arctic char and eel.

The pike is one of our largest freshwater fish, growing to a length of 1.5 m (5 ft) and weighing as much as 25 kg (55 lb). They can easily be identified by their large size, torpedo-like shape and powerful jaws. Pike hunt using an ambush method, lying in wait for long periods until their unsuspecting target floats past, then striking suddenly and effectively.

Their main prey species are smaller freshwater fish, but they also occasionally go for ducklings and water voles. Reports of them grabbing the head of a feeding swan and dragging it beneath the surface are untrue, but they do often prey on their own species – their cannibalistic tendencies mean that young pike need to be especially careful.

Pike courtship is a lively affair: several males will gather around the larger female, following her every move and looking for the opportunity to mate with her. Once she has chosen a suitable male, she will slow down and begin to release her eggs, which are then fertilised by the male.

Another iconic fish, the eel, has a truly bizarre lifecycle. Although eels can be found in many of our rivers, none of them is actually born here. Instead, they hatch thousands of kilometres away in a region of the Atlantic Ocean known as the Sargasso Sea. The tiny baby eels – called elvers – then swim right across the ocean and into our river systems, where they feed and grow to maturity. The process takes several years, and only when they reach adulthood do they leave us and head all the way back to their place of birth, to lay their own eggs.

Unfortunately, eels have undergone a severe decline in recent years, partly as a result of water pollution but mainly because the young elvers are harvested in vast numbers to be sold overseas. This illegal but widespread practice will need to be stopped if we are to save this amazing creature.

Eels may have the most extraordinary lifecycle of Britain's freshwater fish, but in terms of peculiarity it's hard to beat the lamprey. Superficially eel-like in size and shape – long, slender and up to 1 m (3 ft) in length – lampreys are jawless fish with a funnel-shaped mouth. Of all the world's vertebrate species, lampreys are considered one of the oldest and most primitive, having hardly changed for over 300 million years. They have a reputation for latching onto other fish and sucking out their insides, but not all species feed in this way.

Lampreys played a minor but significant part in English history – the twelfth-century king Henry I is said to have died from 'a surfeit of lampreys'. They were also popular at Roman feasts, but today they are rarely eaten and, instead, are used in medical and scientific research, as their simple structure makes them useful as study subjects.

The eel has one of the most remarkable lifecycles of any British creature.

Salmon & Trout

Of all our freshwater fish, two reign supreme: salmon and trout. They are the most sought-after by anglers, and also display some fascinating behaviour and specialised adaptations to life in our lakes and rivers. Anyone who has seen salmon leaping as they head upriver from the sea to spawn knows that they have been privileged to witness one of the greatest feats of endurance shown by any wild creature.

Atlantic salmon, to give the species its full name, spend the majority of their lives at sea, feeding and growing until they reach adulthood. But when the time comes to breed, they undertake an extraordinary journey to the place where they were born, way upstream from the coast. Here, they lay their eggs, and then usually die, exhausted by the battle to get to their destination. The journey upriver usually begins in September or October, with the salmon waiting until the conditions are ideal for leaping before they start. Some of the best salmon rivers in

An Atlantic salmon leaping a waterfall is one of the most extraordinary of all British wildlife spectacles.

Britain include the Tyne in the north of England, once so polluted that virtually nothing could survive there, and the Tay, Dee, Spey and Tweed in Scotland.

As soon as the eggs hatch, the youngsters start to go through several distinct phases of growth for several years, before they head back towards the Atlantic. The early stages are a dangerous time for these tiny fish, as all sorts of predators, including larger fish, herons, kingfishers and otters, feed on them. Only about one in 20 survive to become adults.

Salmon are usually a silvery colour, but males can appear crimson on the belly when they are in breeding condition. They are our largest freshwater fish, with a few individuals growing more than 1 m (3 ft) in length. They need to be big and strong: when leaping, a salmon can reach a height of 3 m (10 ft), which is essential if it is to get over some of the barriers on its way upriver. The record weight for a rod-caught salmon in Britain is 29 kg (64 lb), hooked by Miss Georgina Ballantine on the River Tay in 1922.

In recent years, a major industry to farm salmon has grown up around our coasts, particularly off the west coast of Scotland. Farmed salmon are kept in special pens, which, unfortunately, increases the risk of disease and parasites, which can then spread to the wild population. Escaped farmed salmon may also threaten wild fish by hybridising with them.

Two different forms of trout can be found in Britain, distinguished from one another by their habits and lifecycle. Brown trout are found only in fresh water, while sea trout, as their name suggests, spend part of their lives in the marine environment but, like salmon, venture upstream along rivers in order to breed. Both are medium-sized fish, with dark spots on a silvery-grey background.

Trout are an excellent indicator of the health of a river. They require clean, fresh and oxygenated water (chalk streams are one of their favoured habitats), and plenty of vegetation along the banks, where they can hide from predators. They feed mainly on larger insects and other aquatic invertebrates, which means they can be tricked into taking an angler's hook baited with a carefully tied 'fly' made of feathers and designed to resemble the trout's usual prey. Larger trout will also feed on a wide range of other creatures, including baby birds and water voles.

The brown trout is much sought-after by anglers, who prize this fish for being difficult to catch.

Trout produce an enormous number of eggs – about 2,000 per kilogram (2¼ lb) of the female's body weight – but the vast majority of the young will die long before they reach maturity, having been consumed by a range of freshwater predators.

The population of brown trout found in Loch Ness and other Scottish lochs has developed an unusual habit – of eating one another. These fish – known as 'ferox' trout from the Latin word for fierce – have grown so large that they can no longer survive on invertebrates, which in any case are scarce in these huge lochs. By switching their attention to fish, including some of their own species, they can grow to an enormous size: the British record for a rod-caught ferox trout is more than 14 kg (30 lb). They can also live longer, in some cases more than 20 years.

Sharks

More than 20 different species of shark – including the smaller dogfish
– are regularly seen off Britain's coasts. Some of these, including the
porbeagle and basking sharks, are resident all year round, though
may be seen more often in spring and summer. Others, such as the
blue shark and short-fin mako shark, are seasonal visitors, turning up
each summer on their migratory journeys across the Atlantic, while
a few are rare vagrants, only occasionally found in British waters. Of
all marine creatures, sharks are among the most globally threatened.
Britain's species are no exception, with threats including climate
change, overfishing and pollution.

One big problem faced by sharks is that since the film *Jaws*, all
shark species – no matter whether or not they attack human beings –
are regarded as lethal killers. This has led to one of the most intriguing
British wildlife topics in recent years: whether or not great white
sharks have ever been seen off our coasts.

There have been a number of credible sightings, notably off
Cornwall, in recent years. But despite lurid headlines in the tabloid

*Basking sharks are
the second largest
fish on Earth, and
yet are regularly seen
off our northern and
western coasts.*

newspapers, there has still not been a confirmed record of a British great white. Other large sharks can easily be mistaken for this species, so until a definitive photograph or video footage is obtained, or a specimen is caught or found washed up, then the presence of this legendary predator in our waters must remain unproven.

Other reports of 'killer sharks' off Cornwall include supposed sightings of oceanic white-tipped sharks, but, as with the great whites, this may be a case of mistaken identity. However, as climate change continues to bring species from tropical waters closer to our shores, the chances of one of these 'man-eaters' turning up will increase.

In fact, one monster shark does turn up in British waters: the 12 m (39 ft)-long basking shark, which can weigh up to 7 tonnes. Second only to the whale shark as the world's biggest fish, the basking shark is a regular sight off the west coasts of Britain during the spring and summer months. Despite its huge size and fearsome appearance, the basking shark is entirely harmless, and feeds exclusively on tiny marine creatures known as plankton.

Like so many other large marine creatures, basking sharks have suffered in recent years from a range of problems, including being struck by boats as they feed on or near the surface, and getting tangled up in ropes and nets. Pleasure-boat users and fishing-boat owners – especially in the holiday hotspots of southwest England – have been warned to look out for these mighty fish and take evasive action. Meanwhile, conservationists have tagged the fins of basking sharks to discover more about their mysterious lifecycle, and where they go outside the summer season.

The basking shark feeds by vacuuming up huge amounts of tiny marine creatures known as krill.

Basking sharks can sometimes be glimpsed from land, especially on calm days when they are more easily seen as they come to the surface to feed. By far the best way to see basking sharks is, as with whales, dolphins and porpoises, to take a special boat trip with expert naturalists on board. These run during the holiday season from several ports in Devon, Cornwall, west Wales and western Scotland – all basking shark hotspots.

During hot summers, the number of sightings increases dramatically, as warmer water means more extensive blooms of plankton, on which these giant creatures feed. As climate change leads to a rise in sea temperatures around our coasts, it is likely that sightings of basking sharks will become even more frequent and that the current 'season', which runs from spring through to late summer, may extend throughout the year.

Plants & Flowers

Deciduous Trees

Trees may be big but, paradoxically, they are easy to overlook. While watching birds in trees or walking through a wood in search of fungi or flowering plants, we often ignore the largest living organisms around us. However, a closer study of trees can be just as rewarding as any other branch of nature, if you'll pardon the pun. Whether you are interested in the tree itself or the fruits, leaves and bark it produces, there is a lot we can learn from this important part of the British ecosystem, to give us an insight into many other kinds of plants and animals.

Deciduous trees are those that shed their leaves in autumn in preparation for the hardships of winter. There are about 60 native species found in Britain but there are many others that have been introduced, some of which were brought here centuries ago and have become an integral part of our landscape.

Oak woodlands support a wider range of wildlife than almost any other British habitat.

The horse chestnut, for example, was brought to Britain in the seventeenth century during the reign of James I, while its cousin, the sweet chestnut, has been here even longer. It was originally introduced by the Romans, who harvested the nuts for food, which we still do today, while the fruits of the horse chestnut – the famous conkers – have long been collected by schoolchildren for playing the game of the same name.

Certain trees attain an iconic status because they are inextricably linked with a particular landscape or part of the country. The beech, for example, is favoured in southern Britain, its stronghold, owing to its majestic shape and glorious autumn colours. Beech woods are also easy to walk through because the trees' dense canopy means that other plants do not grow in profusion beneath them.

We used to hold the same affection for the elm, which for centuries typified our lowland farmed landscape and, especially, the English country village. Sadly, however, the spread of Dutch elm disease during the second half of

the twentieth century has virtually destroyed Britain's elms, at least in terms of adult trees. Caused by a fungus, and transferred from tree to tree by a small but lethal beetle, the disease attacks youthful saplings and prevents them reaching maturity.

But of all our deciduous trees, without doubt the best known and best loved is the English oak, also known as the common, or pedunculate, oak (pedunculate refers to the stalk on which the acorns, the fruit of the oak tree, hang). The unmistakable outline of this widespread tree can be seen throughout much of lowland Britain. In recent years, oaks, too, have succumbed to the ravages of a disease – Acute Oak Decline (AOD) – but, fortunately, most appear to be able to survive it. Oaks support a vast array of wild creatures, especially insects and other invertebrates, which in turn provide food for woodland birds.

Other popular native deciduous trees include the ash, whose bunches of seed pods, known as keys, are a classic sight; various kinds of willow, usually found near water; and less well-known species such as the hornbeam, whitebeam and alder. The definition of 'tree' is more flexible than it first appears, with smaller species such as elder, hawthorn and blackthorn also included within this group of woody perennial plants.

A woodland walk in spring is one of the classic British wildlife experiences.

Coniferous Trees

In contrast to the 60 or so native deciduous trees found in Britain, there are just three native conifers, although many other non-native species can be found here. The three 'true Brits' are the Scots pine, yew and juniper.

Of these, perhaps the most iconic is the Scots pine. At one time, much of the highland region of Scotland was blanketed with them, creating the Caledonian pine forest. Today, only a tiny fraction remains, mostly in the area around the Spey Valley and farther east in Deeside.

Even much reduced, these stands of majestic trees are still a very impressive sight, and they are also very important from an ecological point of view. The Scots pine forest continues to provide a home for some of Scotland's special birds, including the capercaillie, the crested tit and, most importantly, Britain's only endemic bird species, the Scottish crossbill. The forests are also home to colonies of northern and Scottish wood ants, which make their huge, domed nests out of millions of pine needles.

Scots pine also occurs in many unexpected locations south of the border, where it has been brought over the years. Here, though, it may be harder to identify, as there are often other species of non-native pine, which may cause confusion. Check out the reddish-brown colour of the tree's bark, its distinctive clumps of needles and, especially, the pear-like shape of the pine cone – a favourite for children to pick up and take home.

In the rest of Britain, you may come across a wide range of introduced conifer species. These range from enormous redwoods and the Douglas fir to smaller trees, such as the Norway spruce (the classic Christmas tree), the Sitka spruce and various kinds of cypresses, firs and cedars.

The larches are another widespread group. Often considered native to these shores, they were in fact introduced here from the seventeenth century. Larches are unusual in that they are both coniferous and deciduous: they produce cones but also shed their needles (really a modified form of leaf) in autumn and winter.

The yew, meanwhile, is a true native, and more than any other tree has a central part to play in our history and folklore. Yews are often found in churchyards: look out for a round, bushy tree with plenty of dense foliage, dark bottle-green in colour, and bright red autumn berries that often look as if they are plastic ornaments stuck onto the tree.

Scots pine forests once covered vast areas of Scotland; the wooded valleys of Speyside are now the best place to see this important habitat.

There are two reasons why yews are so common in churchyards and, indeed, sometimes hard to find elsewhere. They are extraordinarily long-lived – some still-living specimens, such as the famous Fortingall Yew in Perthshire, are thought to be well over 2,000 years old. As a result, the yew has long been associated with pagan religions, and when the early Christians came to build their first churches, they often placed them on pagan sites.

The second, more pragmatic, reason is that the bark, leaves and seeds of the yew are poisonous to livestock, which meant that the trees were often destroyed by farmers. Conversely, yews were perhaps also planted in churchyards to discourage livestock from venturing inside.

Yews do not support a huge amount of wildlife, but goldcrests and coal tits often seek refuge in the dense foliage during cold weather, while larger birds such as thrushes and blackbirds love the berries.

Even though the third native British conifer, the juniper, is the most widespread woody plant on Earth, found right across the northern hemisphere from the Arctic Circle, and south to the warm temperate latitudes of Europe, Asia and North America, it is far less familiar than the Scots pine or the yew. This is partly because it looks more like a shrub than a tree. Often found on windswept hillsides and moorland, the juniper can be identified by its needle-like leaves in groups of three, and purplish-black berries – used, of course, to flavour gin.

In recent years, conifers have had a bad press, as large-scale commercial forestry plantations are neither very attractive nor particularly good for wildlife, but attitudes within the industry are starting to change. More thoughtful planting will create forests with different stages of growth, ideal for birds such as the nightjar.

Yew trees are among the most ancient of all our living things – the oldest date back before the birth of Christ.

Orchids

Imported varieties of orchid are frequently sold in florists, so in many people's eyes they are exotic tropical plants. It may then come as a surprise to discover that Britain is home to more than 50 native species. Furthermore, although some of these are so rare that their whereabouts are kept a closely guarded secret, others can be easily seen.

The appeal of orchids is easy to understand. These are magical flowers, both to look at and because of their complex and fascinating biology. No other group of flowers has quite such bizarre lifecycles, involving everything from looking like a bee in order to attract those insects to pollinate the flower to dispensing with conventional floral characteristics altogether and living as a parasite off underground fungi.

Orchids can be found in a wide range of habitats: woodland (both deciduous and coniferous); grassland, especially on chalk and limestone soils; watery areas such as fens, bogs and marshes; moorland and heathland; and, in more recent years, man-made habitats such as churchyards, railway cuttings and brownfield sites.

Early purple orchids are one of the first in their family to appear, usually in April.

The first to appear is the aptly named early purple orchid, which usually begins to bloom in early April, about the same time as two other iconic flowers of spring: cowslips and bluebells. With its distinctive deep purple flowers on a long stalk, it is easy to identify.

Other spring orchids include the early marsh, early spider and green-winged orchids, while from June onwards the vast majority of other orchid species appear, including the group known as helleborines, which mostly grow in woodland. Another woodland specialist is the bird's-nest orchid, named after its roots, which are said to resemble an untidy bird's nest.

The bird's-nest orchid is one of only three British orchids that has no green leaves, and therefore cannot obtain its nutrients in the usual way, using chlorophyll and sunlight. Instead, it does so by exploiting fungi. As it has no need for sunlight, the bird's-nest orchid is one of the very few plants that can live in beech woods, whose dense canopy prevents light reaching the forest floor. It thrives in damp, warm springs, because drought reduces the amount of fungi for it to feed on.

Other orchids also need fungi, but in a very different way. Their seeds germinate underground, and the seedlings need the nutrients provided by fungi to grow before they eventually appear above ground.

However, the best-known association between orchids and another group of living organisms is the strange relationship between some orchids and insects. Lots of flowers are brightly coloured in order to attract insects to pollinate them, but some orchids take this one step further. They actually mimic a female insect in order to tempt the male to land on them. When he does so, the texture of the flower makes him copulate with it, during which he picks up pollen, which he later deposits onto another orchid.

Of all our orchids, perhaps the two most celebrated are also the two rarest. Indeed, the lady's slipper and ghost orchids have a good claim to be the rarest of all Britain's several thousand flowering plants. The lady's slipper, so-named because its flowers resemble that item of footwear, is confined to a single site in Yorkshire, which is kept so secret that even many lifelong orchid enthusiasts are unaware of its exact location.

The decline of the lady's slipper towards extinction was down to botanists, eager to add it to their collection, so such secrecy is understandable. Fortunately, plans are under way to reintroduce the plant to other sites in the Yorkshire Dales, including one that will be open to the public, so there may be a chance to see this unique and beautiful orchid in the future.

This is not the case for the most elusive of our native orchids, the aptly named ghost orchid. It was only found in the Chilterns, to the northwest of London, and the Welsh Marches (Herefordshire and Shropshire) and was thought to be extinct as a British wild flower until, after an interval of two decades, a single plant was found in Herefordshire in 2009.

Green-winged orchids bloom in profusion in marshy areas.

Hedgerow Flowers & Plants

Our hedgerows have taken a bit of a battering in the past half century or so. Indeed, more than one million miles of hedgerow have been destroyed since the end of World War II, in the quest for ever more efficient and intensive agriculture.

Fortunately, despite this wanton destruction, many still remain. They provide both a place to hide and a place to feed for all kinds of wildlife, including nesting birds, small mammals and insects, such as beetles and butterflies. However, were it not for the complex network of hedgerow plants that forms this very special habitat, none of these creatures would find a home there at all.

Most hedgerows were originally planted to form shelter belts for livestock or to divide fields, and the original makers would often use a range of different woody plants in order to create a suitably dense hedge. Classic hedgerow plants include hawthorn, blackthorn, hazel, field maple, guelder rose and spindle, while the bramble is usually also present.

Sloes (above) are the fruit of the blackthorn, and have been used for generations to make sloe gin.

Hawthorn (below) and red campion (below right) are common and familiar hedgerow plants.

In spring, many hedgerows are covered with white blossom, either from the hawthorn or blackthorn, but which one? The two are, in fact, easy to tell apart because of the way they bloom. Blackthorn blossom appears before the green leaves, while the opposite is true of the hawthorn, whose leaves cover the plant a week or two before its white blossom appears. Blackthorn blossom tends to come first, usually in early April but in warm springs often even earlier.

Hawthorn blossom gave rise to the well-known proverb: 'ne'er cast a clout till May is out'. This warning not to take off your winter coat until 'May is out' has often been misunderstood to refer to the month of May, but in fact it refers to the alternative name for the hawthorn: mayflower. Other names for this ubiquitous hedgerow plant include the bread-and-butter plant, because country folk used to munch on the young leaves to stave off hunger.

Later in the year, the two plants are much easier to tell apart, as their fruit could hardly be

more different: the bright crimson berries of the hawthorn contrast with the deep, bluish-purple sloes of the blackthorn, which are used to make the traditional beverage sloe gin.

Hedgerows also often contain larger trees, either planted many years ago or, more often, trees that were already there when the boundary was originally laid down. Oak trees are probably the most frequently occurring, though in wetter areas, rows of willows are a common sight. The non-native sycamore and the native ash may also be found. These trees add another dimension to the wildlife potential of any hedgerow by providing different 'micro-habitats' where other creatures can flourish.

All sorts of wild flowers thrive in the area of ground next to hedgerows. These include members of the thistle family such as the common knapweed and sowthistles, and the attractive white campion and red campion. Foxgloves are also often found along hedgerows, their tall stems of purple flowers sometimes reaching head height. Another pinkish-purple flower, the great willowherb, is frequently seen in mid- to late summer, as is the non-native but widespread red valerian.

The commonest and most obvious family of hedgerow plants, often found along roadsides in the small area of ground between the hedge and the road itself, is the umbellifers, including hogweed, rough chervil and, best known of all, cow parsley.

As its name suggests, cow parsley (along with its larger cousin hogweed) was often fed to domestic livestock. Its intricate creamy white flowers have also led to the alternative, and rather more pleasing, name of Queen Anne's lace. Cow parsley flowers from April to June, while hogweed appears from May through to August, its long stems supporting clumps of milky white flowers, often covered with tiny insects.

Woodland Flowers

If you want to find a variety of attractive and fascinating flowers, it's hard to beat a walk in an ancient woodland. From early spring through to late summer, there will always be something to see, even if you may have to search a bit more carefully to find the most interesting specimens.

In the milder southern and western parts of England and Wales, especially in sheltered areas near the coast, woodland flowers sometimes appear as early as late January or the start of February, while in the Scottish pine forests you may have to wait until late April or even May.

Candlemas (2 February) is traditionally the date when the first snowdrops appear, though in very mild winters they occasionally bloom before the previous year is over. Although they are revered as one of our national flowers, snowdrops are almost certainly an introduced species. Probably brought here from southern Europe during Tudor times, the snowdrop found our damp, mild winters to its liking and rapidly spread throughout much of lowland Britain. The sight of the first snowdrops on a sunny day in the early part of the year is always a welcome reminder that spring is just around the corner – no wonder we love them so much.

For the eighteenth-century naturalist Gilbert White, author of *The Natural History of Selborne*, the lesser celandine was the true harbinger of spring. This tiny yellow flower usually appears in late February, often in a sunny spot on the edge of a wood or in an open glade. Soon afterwards primroses appear – the name derives from the Latin *prima rosa*, meaning 'first flower' – swiftly followed by wood anemones and wild garlic, or ramsons, whose pungent smell fills the woodland air by Easter.

As spring progresses, other common woodland flowers include violets and forget-me-nots, along with Britain's favourite wild flower, the bluebell. Bluebells first appear in the milder southwest in early April, and during the next few weeks come into bloom in woods up the country until they finally appear in Scotland in late May or early June.

A display of bluebells is one of our most beautiful natural

Snowdrops (above) and bluebells (below) regularly top the poll of Britain's favourite wild flowers.

spectacles, especially when you consider that the species is rather scarce away from our shores, so they are not just attractive but important, too. Unfortunately, in recent years our bluebells have been threatened by the imported Spanish version, whose flowers lack the 'nodding' character of the native variety, and are more open and saucer-shaped in appearance. These escapees from gardens and parks hybridise with the British bluebells, and may need to be controlled in the future to stop them destroying the purity of our favourite native plant.

Another iconic woodland flower is the daffodil – not the gaudy cultivated varieties, with their huge flowers and custard-yellow or orange colours, but the genuine wild daffodil. These can be identified by their trumpet-shaped flowers with pale yellow flowers and a darker yellow trumpet in the centre. The leaves are long and narrow, and greyish-green in shade. Wild daffodils were made famous by the Romantic poet William Wordsworth, who wrote of coming across 'a host of golden daffodils' while on a spring walk in his beloved Lake District.

Other woodland plants are under threat for several reasons. Woods and forests once covered the vast majority of the British landscape, but over the years clearing has reduced woodland cover to about one-tenth of what it once was. Woodland management also affects the number and diversity of plants: many woodlands have been allowed to flourish without regular cutting, which means that less sunlight now reaches the forest floor, making it harder for plants to grow.

Now conservationists are reintroducing the old woodland practices such as coppicing and pollarding, in order to make clearings and rides where pools of light can create the right conditions for these beautiful wild flowers to flourish.

Lesser celandine (left) and primrose (below) are amongst the earliest wild flowers to appear in spring.

Meadow Flowers

Of all our wildlife habitats, none has been quite so relentlessly and effectively destroyed as the traditional hay meadow. For thousands of years after Neolithic settlers first cleared the forests and allowed their animals to graze the land, hay meadows were a classic sight in the lowland British countryside, and remained so until the start of World War II.

But the need to be self-sufficient in food during wartime and in the following decades saw the wholesale destruction of these valuable sites, as they were ploughed up and planted with wheat and barley. Since the end of the war, an estimated 98 per cent of all Britain's traditional hay meadows have been lost, and the few that remain are almost all now nature reserves. Fortunately, under careful management, many are now thriving.

The key to the success of hay meadows as a wild-flower habitat was down to three things: regular grazing by domestic animals such as cattle and horses; the cutting of grass for hay in late summer; and leaving the areas unploughed and unsprayed with pesticides or herbicides. This carefully managed regime meant that the fast-growing grasses we see in bright green fields today were outcompeted by a far wider variety of grasses and wild flowers.

Classic wild flowers of a hay meadow include cowslips and buttercups, which have both given rise to a wealth of folklore. For instance, children know that if you place a buttercup beneath your chin and it shows yellow, you like the taste of butter. An alternative name for the cowslip is 'fairycup', because the drooping bell of the flower is supposed to be the ideal place for a fairy to sit. Shakespeare himself referred to this in *The Tempest*, where Ariel sings, 'Where the bee sucks, there suck I, in a cowslip's bell I lie.'

One of the commonest wild flowers, the daisy, is also linked with the world of magic. Daisy chains, that perennial favourite of country children for centuries, are supposed to protect their wearer against being abducted by fairies, while the very name 'daisy' is a corruption of 'day's eye', as the flowers only open when the sun starts to rise at dawn.

Much rarer is the snake's head fritillary, a stunning little plant that is now mainly known from a few sites in Gloucestershire and Oxfordshire, including the meadows around Oxford. It is so-called because its pinkish-purple flowers, spotted with white, look rather like the head of a snake.

Midsummer meadow flowers include viper's bugloss, meadow cranesbill, lady's bedstraw and bird's foot trefoil, all of which appear from June through to September. Bird's foot trefoil is sometimes called the 'egg and bacon plant', because it has yellow flowers streaked with red. The name 'bird's foot' comes from the very distinctive seed pods, which have three prongs, and which look remarkably like the three toes of a bird.

Another plant that blooms at this time is the yellow rattle, so-named because once the seeds are set, they rattle when the case is shaken. Also known as cockscomb, the yellow rattle is a parasitic plant that feeds off the plants around it, which makes it very useful for conservationists, as it prevents less desirable grasses from spreading in the early days of a hay meadow's restoration.

Later in the summer, the colours in a hay meadow subtly switch from yellows and pale mauves to deeper purples. Thistles and knapweeds appear in profusion, along with darker brown docks. Thistles, in particular, attract large numbers of grassland butterflies, including meadow brown and gatekeeper, as well as the blues and marbled white.

As the summer turns to autumn, after the grass has been cut, the colours finally disappear until the following spring, when a new crop of meadow flowers appears again to brighten the landscape.

Thistles are at their best in late summer, when they provide nectar for bumblebees, butterflies and day-flying moths.

Marsh Plants & Flowers

Marsh marigold (right) and lady's smock (below) are two classic plants of wetlands in early spring.

Since World War II, many of Britain's wildlife habitats have suffered severe losses, with predictable effects on the fortunes of their wildlife. But while the fauna and flora of woodland and farmland continue to suffer, one habitat and its wildlife are making a comeback. Wetlands are reappearing all over lowland Britain, thanks to the enlightened 'rewilding' policies of national and local conservation organisations. As a result, our wetland wildlife is thriving.

Although much of the publicity goes to flagship species such as ospreys or cranes, wetland plants are benefiting too, with many species now extending their range and appearing in greater profusion than for at least a generation.

Late March and early April see the first wetland plants beginning to come into flower. On wet meadows and marshes throughout Britain, two of the first flowers to appear are lady's smock and marsh marigold, whose blooms emerge at about the same time as marshland birds such as the redshank, lapwing and snipe are beginning their courtship displays, heralding the coming of spring to our wetlands.

Lady's smock is one of our most delicate and attractive native plants, with a wealth of folklore associated with it. It is not a showy flower, appearing almost white at a distance, but a close look reveals a delicate pale lilac shade to its petals. Its peculiar name derives from its supposed similarity to the smocks worn by Tudor women.

Perhaps a better name is its alternative, 'cuckoo-flower', given because the flowers usually emerge at roughly the same time as cuckoos return to our shores from their winter quarters in Africa, around the middle of April. Another folk-name, from Hampshire, is 'nightingale-flower', again celebrating the coincidence of first flowering with that iconic bird's return.

Marsh marigold also has a wealth of folk-names: in his seminal work *The Englishman's Flora*, botanist and folklorist Geoffrey Grigson listed at least 80, including 'bachelor's buttons', 'gipsy's money' and 'water blobs'. The appearance of its custard-yellow flowers during spring must have cheered our ancestors after the long winter, and the plant is useful, too: among other things, it has been used as a dye, medicine and food.

From June to August, an even more colourful and impressive plant – the purple loosestrife – joins them. Also known as 'long purples' because of its height, these deep purple flowers bring a new splash of colour to their watery habitats.

But not all marsh plants are so common and widespread. Some, such as the fen orchid and fen violet, are now confined to a few locations; once they were far more widespread, but the wholesale drainage of the fens led to a long decline. Today, the fen orchid is found only in the Norfolk Broads and a dune system in South Wales. The range of the fen violet, meanwhile, is even smaller – it can be found on only three nature reserves: at Woodwalton and Wicken Fens in East Anglia, and at Otmoor in Oxfordshire.

Later in the year, other classic marshland plants make their appearance. Water lilies and yellow flag bloom in ditches, dykes and ponds – anywhere with enough water to sustain them through the heat of the summer. Although they appear to be floating on the water's surface, water lilies do have roots, which anchor them to the bed of a stream or pond. But some floating plants do just that – the various species of duckweed simply rest on the surface of the water, often creating a green blanket and preventing sunlight reaching below.

The ability of duckweed to completely cover the surface of water led to a grisly folk tale in parts of Britain – that of Jenny Greenteeth. This mythical water fairy is supposed to tempt little children to walk across the layer of duckweed then pull them down into the watery depths to their death. Like so many fairy tales, this was presumably designed as a warning to children of the dangers of venturing too far afield.

Purple loosestrife (above) and yellow flag (below) are two of the commonest wetland plants in late summer.

Moorland & Heathland Plants

Vast areas of the north and west of Britain – especially those over a few hundred metres in altitude – have a very different fauna and flora from the lowland areas to the south and east. Indeed, at first sight these huge tracts of moorland appear to be virtually lifeless but, as with all habitats in upland areas, you just need to take a closer look to discover a wealth of fascinating wildlife.

Although moorland may look like a quintessential 'natural' landscape, it is carefully managed. Without human intervention, often over many centuries, this land would still be covered with trees. Our distant ancestors burned the forests to create grazing areas for sheep, and, in the past couple of centuries, large areas of moorland, especially in northern England and Scotland, have been managed for the shooting of one particular bird: red grouse.

The key plant on grouse moors is heather. Its young shoots provide food, while the more mature plants give the birds a place to shelter and hide from predators and, of course, shotguns.

The plant we call 'heather' is in fact just one of a whole family of hardy evergreen shrubs, including bell heather and cross-leaved heath. Most of them have the same pinkish-purple flowers, though, occasionally, heather produces white blooms, traditionally celebrated as a bringer of good luck. Also known as 'ling', heather has long had a wide range of practical uses, including as roofing material, fuel, and food for sheep. Its flowers also produce a wonderfully dark, rich and full-flavoured honey.

Other classic moorland plants include crowberry and bilberry, both of which provide food for birds and small mammals. Wetter areas and bogs will often be covered with sphagnum moss. This extraordinary plant has the ability to soak up vast amounts of water, and for this reason was commonly used as a dressing for wounds during World War I.

The higher upland areas are home to all sorts of rare plants, many of them relics left over from the alpine flora of the last Ice Age. These include the tufted saxifrage and Snowdon lily in North Wales. Both are just managing to cling on in their north-facing mountain homes in the face of the growing threat of climate change. Scotland is also home to many scarce alpine plants, including the mountain avens and alpine sowthistle, the latter now confined to just a handful of sites.

Heather is the dominant plant of many moors and heaths, especially in Scotland and northern England.

A typical scene on an upland moor, dominated by bilberry and heather.

Way to the south, another group of plants exploits a similar habitat: lowland heath. Like upland moors, heaths are on poor soils and suffer extremes of climate – in this case, though, heat rather than cold. Lowland heath is mainly found in the southern counties of Dorset, Hampshire and Surrey, with some areas also in Devon, Cornwall and East Anglia.

As in the uplands, common heather is one of the dominant plants, but alongside it, and often dominating the landscape with its bright yellow flowers, is the classic plant of lowland heath: gorse. Gorse can be found throughout most of Britain, but it comes into its own on these dry, sandy soils. Its success may well be down to the fact that it is able to flower all year round – there is a saying that 'when gorse is in flower, kissing is in season', a wry comment on the plant's ability to bloom every month. Alternative names for gorse include 'furze' and, in the northern and eastern parts of Britain, 'whin', which gives its name to one of the classic birds of heathland and moorland, the whinchat.

In mid- to late summer, the gorse produces seeds in hard, black pods, which then go on to burst during warm weather in July and August, producing a surprisingly loud crack as they do so. On southern lowland heaths, areas of heather and gorse are often interspersed with belts of trees, including Scots pine and birch, which must be artificially controlled by cutting down or burning if the heathland is to be kept as an open landscape.

Urban & Roadside Plants

All our flowering plants evolved at a time when there were far fewer people living in Britain, and the landscape had yet to be covered in the concrete required to support so many millions of inhabitants. As a result of this, many species have struggled to survive during the past two centuries of rapid urbanisation and the growth of towns and cities. But not all our plant species have declined. Some, including once rare and specialised species, have thrived in our urbanised society, and today are among our commonest, most widespread and most familiar plants.

Perhaps the most familiar plant of urban waste ground and other open spaces is the rosebay willowherb. Known in its native North America as 'fireweed', for its ability to colonise areas of land that have been recently burned, this attractive plant with its tall stems covered in pinkish-purple flowers is a familiar sight in the summer months, often along the verges of roads. Yet, until the middle of the twentieth century, it was quite a scarce plant in Britain. Its big opportunity came, ironically, in the decades following World War II, when it colonised the many bomb sites dotting the urban landscape. Having gained a foothold in our cities, it then began to spread throughout Britain via roads and railway lines, with the plants readily colonising the verges and cuttings ahead of other, less opportunistic, plants.

The pinkish-purple flowers of rosebay willowherb bring a welcome splash of colour to our roadsides each summer.

The tale of another common plant of urban and brownfield sites is even more extraordinary. The Oxford ragwort is so-named because the first specimens brought to Britain during the early decades of the eighteenth century were planted in Oxford University's Botanic Garden. The plant's only known native home was the volcanic slopes of Mount Etna, in Sicily – a factor that was the key to its later success.

More than 100 years later, in the middle of Queen Victoria's reign, the railway finally came to Oxford. Soon afterwards, some feathery seeds from the Oxford ragwort drifted on the breeze towards the station, where they were caught in the slipstream of a passing train and blown into one of the railway carriages. During the course of the train's journey, as it slowed down and speeded up, the seeds blew out of the carriages and landed on the hard clinker by the side of the railway track. Coincidentally, this was almost exactly the same as their native habitat on the rocky slopes of the volcano, and the seeds germinated and grew into tall plants, with sunshine-yellow flowers. In the 150 years or so since, the Oxford ragwort has managed to spread throughout much of the British Isles, colonising brownfield sites and patches of waste ground in all our urban areas.

Buddleia is also known as the 'butterfly bush' as it acts as a magnet for nectar-loving insects.

Another classic plant of towns and cities has made an even longer journey to get here. Buddleia is originally native to the foothills of the Himalayas, where it grows on rocky slopes in harsh conditions. Having been brought to Britain by keen gardeners in about 1890, the buddleia soon escaped from the confines of parks and gardens and found an ideal home on walls and the sides of buildings throughout urban Britain. Its huge purple flowers are ideal for insects, especially butterflies in search of nectar – hence its familiar name of 'butterfly bush'.

Rosebay willowherb, Oxford ragwort and buddleia are all fairly obvious plants, especially when they bloom during the summer months. But another group of urban plants is easy to overlook, paradoxically because of their huge size. Britain's cities are well planted with trees, especially drought-tolerant species such as the London plane, which provides shade for city dwellers. Most importantly, though, they create a green environment in some of our most populated urban spaces, reminding us that the countryside is not all that far away.

Grasses, Sedges & Rushes

Grasses, sedges and rushes are so familiar and ubiquitous that we tend to take them for granted. Their similarity to one another, and the fact that they lack the colourful blooms of many of our other wild flowers, also means that these groups of plants tend to appeal to the specialist rather than the beginner. This is a pity, as they have their own fascinating stories to tell. But before we uncover some of them, what are the fundamental similarities and differences between these three apparently similar groups?

Despite first appearances, grasses, sedges and rushes are all flowering plants but, because they are pollinated by the wind, they have no need for showy flowers to attract insects. Instead, they produce very light pollen, which can easily be carried away by a passing breeze.

Grasses are one of the most common and widespread of all groups of flowering plants, with about 10,000 species found globally, of which 150 or so are either native to, or naturalised in, Britain. Grasses, of course, play a vital role in feeding human beings and their livestock.

Marram grass is a vital part of dune ecosystems as it fixes the sand and stops it blowing away in strong winds.

Wheat, barley and oats are all grasses that have been modified to produce harvests of grain, while grass is also grown extensively as hay or silage to feed domestic animals such as cattle.

Sedges are also very widespread globally, with more than 5,000 different species. Whereas grasses can grow in a wide range of habitats – dry and wet, highland or lowland – sedges are usually associated with damper habitats, either in the shade of woodland or, most commonly, on boggy ground in wetlands.

One of the most widespread grass species is also one of the tallest. The common reed (often known by its scientific name *Phragmites*) grows in and around almost any area of standing water, and forms the extensive reed beds in many of our wetland nature reserves. In the past, reeds were commonly harvested to be used in thatching roofs or making baskets and other products. Having fallen into decline, these traditional industries are now making a comeback, providing a much-needed boost in some rural economies.

Reeds are also an essential part of the wetland ecosystem of wildlife-rich areas such as the Norfolk Broads and the Somerset Levels. They provide a home to a wide range of waterbirds, including the reed and sedge warblers and the reed bunting, as well as other less expected creatures such as the harvest mouse, which weaves its tiny nest from grasses suspended from the stems of reeds.

Telling grasses and sedges apart can appear tricky at first, but there is one useful distinction: 'sedges have edges' – whereas the stems of grasses are round and hollow, those of sedges are triangular in cross section, giving them a more substantial and solid appearance and feel. Their leaves are also different: sedge leaves are arranged in ranks of three, while those of grasses are arranged alternately. The many species of sedge can also vary considerably in appearance. Despite its common name, cotton grass – which appears in white fluffy patches across boggy areas in summer – is in fact a kind of sedge.

Rushes are superficially similar to grasses and sedges, and also tend to grow in wet or boggy habitats. The best-known genus, *Juncus*, contains several widespread species, which have hairless leaves. In contrast, their relatives, the woodrushes, have long, white hairs on their surface. In medieval times, rushes were often used as floor material.

Although not technically a rush, the bulrush, also known as reed mace, is easily distinguished from other aquatic plants by its long, tall stems topped with a clump of flower heads, which form a sausage-like spike at the top of the plant.

Cotton grass – actually a kind of sedge – is a common sight on damp, boggy areas of moorland in northern Britain.

Berry-bearing Plants

Plants have a problem: they must reproduce and spread to new areas, yet, unlike birds, insects and mammals, they are quite literally rooted to the ground and cannot move. They must therefore rely on other ways to spread their seeds. One of the most ingenious of these requires the production of colourful and fleshy fruit such as berries.

Holly berries appear late in the year, making them a vital food resource for hungry birds.

Technically, berries and fruit are the same thing, but we tend to use the word berry to refer to smaller fruits. A berry is, quite simply, a way of tempting a wild creature to spread the seed within. This goal is achieved by covering the seed in soft flesh, which is nutritious to a wide range of birds and mammals – including, of course, us!

The fruit may be eaten, but the hard seed can withstand its passage through a bird or mammal's gut and emerge when it defecates. So, by eating the food and later depositing the hard seed elsewhere, other creatures unwittingly facilitate the spread of the plant from which it came.

Berries are colourful, too: generally either deep blackish-purple (blackberries, elderberries and sloes, for example) or bright red (hawthorn). This is to appeal to the visual sense of birds, for which these colours stand out, making the berries easier to find. Generally, it is thought that evergreen plants produce red berries, as these show up better against the dark green foliage, while plants with leaves that turn brown in autumn produce black berries, again because they are easier to see.

Berries are very much an autumnal phenomenon, their production coinciding with a period during which birds and small mammals must fatten up, either to survive the coming winter, or to enable them to make their long migratory journeys south.

Long-distance migrants in particular, which must travel thousands of kilometres across the Mediterranean Sea and Sahara Desert to equatorial and tropical Africa, switch their diet at this time of year to take advantage of the bumper harvest. In late August and early September, species such as the whitethroat, lesser whitethroat and garden warbler head to hedgerows and gardens in search of the bright red hawthorn berries and the deep blackish-purple fruit of the elder. A single bird can consume hundreds of berries in a single day, enabling it almost to double its body weight before heading off on its global travels.

Later in the autumn, long after these summer visitors have left our shores, a new group of birds comes to take advantage of this free and abundant food supply and our mild winter climate. From October onwards, winter thrushes such as fieldfares and redwings arrive here in their millions from Scandinavia and Iceland. They may even be joined by a much rarer visitor, the waxwing, flocks of which can be seen feeding on red berries in our urban gardens throughout the winter. Berries are a vital part of the winter diet for all these birds, especially when the ground is hardened by cold or covered with a layer of snow, and other sources of food are hard to find.

Guelder rose (above) and bramble (below) produce plenty of berries, a welcome feast for birds in autumn.

We use berries too, of course. Every child has surely at some time been blackberry picking, while sloes – the hard, purple fruit of the blackthorn – are the key ingredient in sloe gin, and elderberries can be made into wine. But not all berries are edible: even though the birds find them tasty, some, such as holly berries, are actually toxic to humans.

Berry bushes are an excellent way of attracting birds to your garden. Native varieties such as holly, ivy, elder and hawthorn work best, but even exotic garden plants such as cotoneaster and pyracantha attract plenty of hungry birds. Don't clear up fruit or berries that have fallen onto the ground – they are a useful food source for all sorts of small mammals, including hedgehogs, voles and squirrels, as well as late autumn butterflies such as the red admiral, which feed on the juices of fruits as they begin to rot.

Carnivorous Plants

The very idea of a carnivorous plant seems somehow against the laws of nature, yet a significant minority of the world's plants – more than 600 different species in all – has evolved to feed on insects and other invertebrates. Of these, only a handful occur in the wild in Britain, but they are worth seeking out to witness their extraordinary lifestyle.

The reason plants adapt to become carnivores is simple: it is more efficient for them to obtain their nutrients in this way than through conventional means. Carnivorous plants tend to occur in particular habitats, usually where the soil is thin and lacking in essential nutrients. In Britain, carnivorous plants are usually found on acidic soils, which do not support a wide range of other vegetation, notably on moorland and heathland, or in bogs.

The best-known carnivorous plant found in Britain is undoubtedly the sundew. There are, in fact, three different species of sundew found here: the round-leaved, great and oblong-leaved, of which the round-leaved is by far the most common and widespread. All three appear very similar to the untrained eye: a few centimetres tall, with greenish-yellow leaves, each of which is covered with tiny red hairs.

The sundew's trapping method is simple but highly effective. A fly or other small insect is attracted by the sweet-smelling sticky substance (the 'dew') produced by the plant, and lands on the surface in search of a free meal. But as soon as it does so, these sticky secretions, which surround the hairs, trap the insect. As it struggles, the leaves fold over, trapping it even more effectively. In a short time, the plant's enzymes begin to digest the insect, and its leaves absorb the resulting liquid.

Sundews are found on boggy areas of heathland or moorland, often in association with another classic bog plant, sphagnum moss. Classic sites include the Dorset heaths, the New Forest, Dartmoor, Exmoor and the North York Moors. The plant is also fairly widespread in Scotland and parts of Wales.

The two native British species of butterwort – common and pale – are also carnivorous, and use a similar strategy to the sundew to catch their prey.

Butterwort may look pretty, but it is a carnivorous plant, which survives by catching small insects.

Both have a rosette of leaves around their base, roughly the shape and appearance of a starfish, onto which they exude a sticky substance, resembling water droplets, which attracts the insects. Again, once the creature lands, it finds it is unable to escape, and death is inevitable.

Not all carnivorous plants live a terrestrial existence. Another family, the bladderworts, are floating aquatic plants, usually found on the surface of ponds, pools and ditches. As their name suggests, their leaves contain tiny bladders, which they use to trap minute water creatures.

The peculiar habits of these carnivorous plants led our ancestors to form some strange beliefs about their supposed magical powers. As its 'dew' never dried out, even in the midday sun, it was thought that the sundew could cure consumption and whooping cough, while the sticky substance was rubbed onto the skin to protect against sunburn. Sundew was also considered to work as a love-charm, attracting the opposite sex, but once our ancestors finally discovered the plant's carnivorous habits, they soon changed their minds!

Butterwort was also supposed to have magical powers. The Scots believed that picking the plant gave protection from witches, while if a mother of a newborn baby drank milk from a cow that had grazed on butterwort, the baby would never be taken away by fairies.

The sticky substance coating the leaves of the sundew may look attractive, but it is a lethal trap for unsuspecting insects.

Ferns, Mosses and Lichens

In any woodland, only a fraction of the plant life you see belongs to a species of conventional flowering plant. Much of the woodland flora belongs to more primitive and unusual groups such as ferns, mosses and liverworts.

Superficially, ferns are quite similar to other plants: they have roots, stems and leaves, for example, and, unlike mosses, they are classified as vascular plants. But the resemblance ends there, for ferns do not produce seeds or flowers. Instead, they reproduce through spores.

Globally, there are about 12,000 different kinds of fern, though only about 60 or so occur native in the wild in Britain. By far the most common and widespread of these is bracken, a large and distinctive fern, easily identified by its huge, triangular fronds, which, in turn, are divided up into a fan-like pattern. Brackens are one of the oldest and also one of the most successful groups of plants, having colonised every continent apart from Antarctica.

In Britain, bracken is often found on the sides of hills or open moorland, and also on the edge of woodland and farmland. Its rapid rate of reproduction makes it something of a problem for land managers and farmers, as it can cover an area very rapidly indeed – in the past, it has sometimes been the focus of official eradication programmes to stop it spreading out of control.

Other ferns found commonly in Britain include the rusty-back fern and the maidenhair spleenwort, which both grow on the shady side of dry-stone walls. The rusty-back is so-named for the colour of the underside of its leaves, while the spleenwort has long fronds, which are said to resemble the tresses of a young woman's hair.

Ferns attract their fair share of enthusiasts. Known as pteridologists, they founded their own group, the British Pteridological Society, in the Lake District in 1891. More than a century later, it is still going strong.

Mosses perhaps have fewer friends, though they too have their own organisation, the British Bryological Society, dedicated to the closer study of the group known as bryophytes, which includes mosses and liverworts. These plants are far more primitive than ferns, with a heritage dating back 450 million years. Mosses and one group of liverworts do have stems and leaves, but no other characteristics of more advanced plants. Another group of liverworts does not show a

Bracken is one of the most widespread of all ferns, often covering the ground in woodland and on the edge of moorland.

difference between stems and leaves, and simply appears to be a small strip of green.

Mosses and liverworts all prefer damp conditions – they are unable to move water around themselves and so act like a sponge. As a result, they thrive on the shady side of buildings and, especially, large, mature trees. Like ferns and fungi, they reproduce by means of tiny spores, distributing vast numbers of them into the air.

Lichens were once classified as primitive plants but are now known to be a symbiotic partnership between a fungus and an alga, in which both organisms help one another to survive. They are one of the most bizarre of all natural phenomena and, like ferns, liverworts and mosses, reward close study. Incredibly long-lived – a single colony of lichens may be many centuries old – lichens are also incredibly tough, able to withstand freezing temperatures and heat waves that would kill most other organisms.

Lichens also prefer cool, damp surfaces away from the glare of the sun. For this reason, they are commoner in the west of England than the east, preferring the damp, mild, maritime climate of west Wales, Devon and Cornwall, where the most species can be found.

In some ways, lichens are so ubiquitous that we tend to overlook them. But take a closer look with a magnifying glass at some of the more unusual specimens and you will discover a hidden beauty.

Like all mosses, maidenhair moss prefers damp, shaded places.

Fungi

Fungi were once considered to be lower forms of plant life, but in recent years it has been discovered just how different they are from other organisms, and they have been given their own 'kingdom'. This puts them on a par with both plants and animals. Indeed, recent genetic studies suggest that fungi may even be more closely related to animals than plants, though they are very different from either group.

The fungi kingdom also encompasses moulds, yeasts and mildews. From a natural history point of view, it is the conspicuous fruiting bodies, known as mushrooms and toadstools, that are of the most interest. Yet these are merely the visible fruit of what are often vast and complex organisms, which may stretch for huge distances beneath the soil.

Fungi perform several vital functions, with the most important being to aid decomposition. Unable to obtain energy by feeding, like animals, or by photosynthesis, like plants, they instead do so by digesting dead or decaying matter in the soil. Living their lives underground, they only emerge as mushrooms in order to disperse their spores – either via the wind or by being eaten and then excreted by a bird or mammal.

In this way, they play a vital role in specific ecosystems and also for the whole web of life on Earth – without fungi, it is no exaggeration to say that nothing else could survive. So, although some species of fungi are parasitic, living off plants without giving anything back, others have a symbiotic relationship with their partner, giving something in return. For example, many fungi grow on the roots of trees, but as well as absorbing the nutrients they need from the roots, they also pass others back to the tree. In addition, the seeds of some plants, including orchids, actually depend on the fungus in the surrounding soil to germinate in the first place.

Mushrooms – the tasty, nutritious fruiting bodies that the fungi produce – have long been sought-after as a food for humans and, indeed, other large mammals. But there is a sting in the tail: some may look tasty but contain a lethal poison. While you can gather mushrooms to eat, you really do have to be very careful that they are edible ones. The situation is made more complicated because some poisonous fungi closely resemble perfectly safe ones. If you are in any doubt at all, leave well alone!

By far the safest and most environmentally friendly way to forage for mushrooms is to take part in a 'fungal foray' with an experienced expert, who will be able to give you reliable advice about which are safe to eat, as well as telling you some fascinating facts. Fungal forays are run in many parts of the country, by local experts and conservation organisations, and usually during the peak fungi season of September through to November. Mild, damp weather is best – frost can kill off the fruiting bodies very quickly, while prolonged drought means they will often not appear in the first place. But fungi are very unpredictable, and may appear almost anywhere – including your back lawn!

Search your local woodland, especially if there are plenty of dead or dying trees, as many species grow on dead wood. Check out the different shapes and colours, and also some of the weirdest names in nature, including yellow brain fungus, brown roll-rim, King Alfred's cake, stinking russula and the sickener! The stories behind these names are equally fascinating. King Alfred's cakes, for instance, are so-called because, when cut open, their insides resemble a charred morsel of cake, alluding to the tale of the Anglo-Saxon monarch who left a batch of cakes to burn to a cinder.

Dryad's saddle is a bracket fungus, growing on dead wood. It is also known as pheasant's back mushroom because of its mottled shading.

Index

Entries in *italics* indicate photographs.

Acknowledgements

Thanks go to everyone involved in *Springwatch* and *Autumnwatch*: the television and online production teams, crews and presenters, the partner organisations outside the BBC, and of course the loyal audience who have enjoyed the programmes over the past few years.

Special thanks go to two of my former colleagues, Brett Westwood and Joanne Stevens, who kindly read through portions of the text and made some very helpful comments.

I should also like to thank Myles Archibald and Julia Koppitz at HarperCollins, together with Helen Coultas who did a splendidly thorough job on the copy-editing; and Ailish Heneberry at the BBC Natural History Unit and Daniel Mirzoeff from the BBC's Commercial Agency for commissioning me to write the book in the first place.

Finally I'd like to say a special thanks to the man who, more than any other, made *Springwatch* the success it is today – Executive Producer Tim Scoones. His vision, commitment and sheer hard work have – along with the efforts of many other people – put British wildlife firmly back at the heart of our TV schedules.

Picture Credits

Top=t; middle=m; bottom=b; Left=l; Right=r

Front Cover ©BBC (presenters) & FLPA (animals); p2 ©Simon Litten/FLPA; pp4-5 ©Bill Coster/FLPA ; p7 ©Marko König/Imagebroker/FLPA; Pp8-9 ©Horst Jegen/Imagebroker/FLPA; p10t ©Paul Hobson/FLPA; p10b ©Herbert Kehrer/Imagebroker/FLPAp; 11tl ©FranzChristophRobi/Imagebroker/FLPA; p11tr ©Konrad Wothe/Minden Pictures/FLPA; p11b ©Franz Christoph Robi/Imagebroker/FLPA; p12t: ©Neil Bowman/FLPA; p12b ©Hugh Clark/FLPA; p13tl ©Erica Olsen/FLPA; p13tr ©Photo Researchers/FLPA; p13br ©John Hawkins/FLPA; p13bl ©Paul Hobson/FLPA; p14tl ©Do Van Dijck/Minden Pictures/FLPA; p14tr ©Erica Olsen/FLPA; p14b © Richard Brooks/FLPA; p15t ©Marcel van Kammen/Minden Pictures/FLPA; p15b ©Andrew Parkinson/FLPA; p17 ©Bill Coster/FLPA; p18 ©John Hawkins/FLPA; p19 ©Steve Young/FLPA; p20t ©Erica Olsen/FLPA; p20b ©Mike Lane/FLPA; p21 ©Gianpiero Ferrari/FLPA; p22t ©John Eveson/FLPA; p22bl ©Andrew Parkinson/FLPA; p22br ©Winfried Schäfer/Imagebroker/FLPA; p23t ©Hugh Clark/FLPA; p23m ©Frans Van Boxtel/FN/Minden/FLPA; p23b ©Mike Lane/FLPA; p24 ©ImageBroker/Imagebroker/FLPA; p25 ©Malcolm Schuyl/FLPA; pp26-27 ©Erica Olsen/FLPA; p28t ©ImageBroker/Imagebroker/FLPA; p28b ©Erica Olsen/FLPA; p29 ©ImageBroker/Imagebroker/FLPA; p30 ©Paul Sawer/FLPA; p31 ©Paul Hobson/FLPA; p32t ©Mike Lane/FLPA; p32b ©Jurgen & Christine Sohns/FLPA; p33 ©Roger Tidman/FLPA; p34t ©Derek Middleton/FLPA; p34b ©Mike Lane/FLPA; p35 ©Derek Middleton/FLPA; p36 ©Robin Chittenden/FLPA; p39 ©Mike Lane/FLPA; p41 ©John Hawkins/FLPA;p42 ©Dickie Duckett/FLPA; p43 ©Duncan Usher/Minden Pictures/FLPA; p44tl ©Bill Coster/FLPA; p44tr ©Robin Chittenden/FLPA; p44b ©Paul Sawer/FLPA; p45t:©Derek Middleton/FLPA; p45b ©Neil Bowman/FLPA; p47 ©Jan Van Arkel/Minden Pictures/FLPA; p48tl ©Robin Chittenden/FLPA; p48tr ©John Hawkins/FLPA; p48b ©Neil Bowman/FLPA; p49t ©Derek Middleton/FLPA; p49m ©Derek Middleton/FLPA; p49b ©David Tipling/FLPA; p50t ©Sean Hunter/FLPA; p50b ©Ramon Navarro/Minden Pictures/FLPA; p51t ©Jules Cox/FLPA; p51b ©Sean Hunter/FLPA; p53 ©Phil McLean/FLPA; p54t ©Phil McLean/FLPA; p54b: ©Tony Hamblin/FLPA; p55t ©David Tipling/FLPA; p55m ©Erica Olsen/FLPA; p55b ©Neil Bowman/FLPA; p56 ©Harry Fiolet/FN/Minden/FLPA; p57 ©Dickie Duckett/FLPA; p59: ©Robert Canis/FLPA; p60 ©Mike Powles/FLPA; p61 ©Martin B Withers/FLPA; P62 ©Martin B Withers/FLPA; p63 ©Simon Litten/FLPA; p64 ©Mark Sisson/FLPA; p65t ©Andrew Parkinson/FLPA; p65b ©Andrew Parkinson/FLPA; pp66-67 ©Andrew Parkinson/FLPA; p68t ©Bill Coster/FLPA; p68b ©Richard Brooks/FLPA; p69t ©Michael Durham/FLPA; p69b ©Jurgen & Christine Sohns/FLPA; p70 ©Marcus Siebert/Imagebroker/FLPA; p72t ©Paul Hobson/FLPA; p72b ©Otto Plantema/Minden Pictures/FLPA; p73tl ©Tony Hamblin/FLPA; p73tr ©Ben Van Den Brink/FN/Minden/FLPA; p73b ©Dietmar Nill/Minden Pictures/FLPA; p75 ©Jonathan Carlile/Imagebroker/FLPA; p76 ©Scott Linstead/Minden Pictures/FLPA; p77 ©Tony Hamblin/FLPA; p78 ©Mike Lane/FLPA; p80 ©Paul Hobson/FLPA; p81t ©Franz Christoph Robi/Imagebroker/FLPA; p81b ©David Tipling/FLPA; pp82-83 ©ImageBroker/Imagebroker/FLPA; p84 ©David Tipling/FLPA; p85t ©Dickie Duckett/FLPA; p85b ©Atlantic Puffin (Fratercula arctica); p86t ©Paul Sawer/FLPA; p86b ©Paul Sawer/FLPA; p86t ©Herring Gull (Larus argentatus); p87m ©Bill Coster/FLPA; p87bl ©Jasper Doest/Minden Pictures/FLPA; p87br ©Mike Lane/FLPA; p88 ©Jose Antonio Moreno/Imagebroker/FLPA; p89t ©David Hosking/FLPA; p89b ©Robin Chittenden/FLPA; p90 ©David Tipling/FLPA; p91 ©David Tipling/FLPA; p92 ©Malcolm Schuyl/FLPA; p93t ©Bill Coster/FLPA; p93b ©ImageBroker/Imagebroker/FLPA; p95 ©Mike Lane/FLPA; p96t ©Bill Coster/FLPA; p96b ©Terry Andrewartha/FLPA; p97t ©Bill Coster/FLPA; p97m ©Marcel van Kammen/Minden Pictures/FLPA; p97b ©Derek Middleton/FLPA; p99 ©Paul Sawer/FLPA; p100 ©Dickie Duckett/FLPA; p101 ©Gary K Smith/FLPA; p102 ©Bob Gibbons/FLPA; p103t ©Paul Sawer/FLPA; p103b ©Jasper Doest/Minden Pictures/FLPA; p104t ©Mike Lane/FLPA; p104b ©Derek Middleton/FLPA; p105 ©Tony Hamblin/FLPA; pp106-107 ©Tony Hamblin/FLPA; p108m ©Paul Sawer/FLPA; p108b ©Erica Olsen/FLPA; p109t ©Jan Sleurink/Minden Pictures/FLPA; p109m ©John Holmes/FLPA; p109b ©Do Van Dijck/FN/Minden/FLPA; p110t ©Dickie Duckett/FLPA; p110b ©Jasper Doest/Minden Pictures/FLPA; p111t ©David Tipling/FLPA; p111m ©David Tipling/FLPA; p111b ©Tony Hamblin/FLPA; p112t ©Chris Brignell/FLPA; p112m ©Dickie Duckett/FLPA; p112b ©David Hosking/FLPA; p113t ©Robin Chittenden/FLPA; p113b ©David Tipling/FLPA; p114 ©Tony Hamblin/FLPA; p115 ©Terry Whittaker/FLPA; pp116-117 ©Guenter Fischer/Imagebroker/FLPA; p118 ©Steve Young/FLPA; p119 ©Roger Tidman/FLPA; p120 ©David Tipling/FLPA; p121 ©Sean Hunter/FLPA; p122 ©Mike Lane/FLPA; p123 ©Malcolm Schuyl/FLPA; p124t ©Stefan Huwiler/Imagebroker/FLPA; p124m ©Horst Jegen/Imagebroker/FLPA; p124b ©Erica Olsen/FLPA; p125t ©Roger Tidman/FLPA; p125b ©David Hosking/FLPA; p126 ©ImageBroker/Imagebroker/FLPA; p127 ©Mike Lane/FLPA; p128t ©ImageBroker/Imagebroker/FLPA; p128b ©Simon Litten/FLPA; p129t ©Winfried Wisniewski/FLPA; p129m ©Mike Lane/FLPA; p129b ©Roger Powell/FN/Minden/FLPA; Pp130-131 ©David Pattyn/Minden Pictures/FLPA; p132 ©Simon Litten/FLPA; p133 ©Sean Hunter/FLPA; p135 ©Elliott Neep/FLPA; p136 ©Derek Middleton/FLPA; p137 ©Tony Hamblin/FLPA; p138 ©Gary K Smith/FLPA; p139 ©Bill Coster/FLPA; p141 ©Paul Sawer/FLPA; p142 ©Sean Hunter/FLPA; p143 ©Erica Olsen/FLPA; pp144-145 ©Richard Costin/FLPA; p146 ©Marcel van Kammen/Minden Pictures/FLPA; p147t ©Paul Hobson/FLPA; p147b ©Jan Vink/Minden Pictures/FLPA; p148 ©Paul Hobson/FLPA; p149 ©Terry Whittaker/FLPA; p151 ©David Hosking/FLPA; p152t ©Mike Lane/FLPA; p152b ©Bert Pijs/Minden Pictures/FLPA; p153t ©House Mouse (Mus musculus); p153m ©Derek Middleton/FLPA; p153bl ©Derek Middleton/FLPA; p153br ©Mike Lane/FLPA; p155 ©Paul Hobson/FLPA; p156 ©Edible dormouse, foraging; p157 ©ImageBroker/Imagebroker/FLPA; p159 ©David Tipling/FLPA; p160 ©Paul Hobson/FLPA; p162 ©Mark Sisson/FLPA; p163 ©Mike Lane/FLPA; p164 ©Paul Sawer/FLPA; p165 ©Jules Cox/FLPA; p167 ©Hugh Clark/FLPA; p169 ©Terry Whittaker/FLPA; pp170-171 ©Terry Whittaker/FLPA; Pp172-173 ©Imagebroker/FLPA; p174t ©Peter Entwistle/FLPA; p174b ©Peter Entwistle/FLPA; p175t ©Peter Entwistle/FLPA; p175m ©Tony Hamblin/FLPA; p175b ©Gianpiero Ferrari/FLPA; p176t ©Derek Middleton/FLPA; p176b ©Matt Cole/FLPA; p177t ©Gianpiero Ferrari/FLPA; p177m ©Gianpiero Ferrari/FLPA; p177b ©David Hosking/FLPA; p178 ©ImageBroker/Imagebroker/FLPA; p179 ©Gianpiero Ferrari/FLPA; p180 ©Gianpiero Ferrari/FLPA; p181 ©Derek Middleton/FLPA; p183 ©Andrew Bailey/FLPA; p184tl ©Matt Cole/FLPA; p184tr ©Gianpiero Ferrari/FLPA; p184b ©Gianpiero Ferrari/FLPA; p185t ©Derek Middleton/FLPA; p185b ©Gianpiero Ferrari/FLPA; p186t ©Martin B Withers/FLPA; p186m ©Ingo Schulz/Imagebroker/FLPA; p186b ©Matt Cole/FLPA; p187t ©Gianpiero Ferrari/FLPA; p187b ©Jef Meul/Minden Pictures/FLPA; p189 ©Matt Cole/FLPA; p190 ©Nigel Cattlin/FLPA; p191 ©Richard Becker/FLPA; p192 ©Malcolm Schuyl/FLPA; p193 ©Chris Mattison/FLPA; p194t ©Steve Trewhella/FLPA; p194b ©Gerard Lacz/FLPA; p195t ©B. Borrell Casals/FLPA; p195b ©Foto Natura Stock/FLPA; Pp196-197 ©Malcolm Schuyl/FLPA; p198 ©Michiel Schaap/Minden Pictures/FLPA; p199t ©Jules Cox/FLPA; p199b ©Gianpiero Ferrari/FLPA; p200 ©Derek Middleton/FLPA; p201t ©Derek Middleton/FLPA; p201b ©Chris Mattison/FLPA; p202 ©Derek Middleton/FLPA; p203 ©Paul Hobson/FLPA; p204t ©Michael Krabs/Imagebroker/FLPA; p204b ©John Eveson/FLPA; p205t ©Ralph Deseniß/Imagebroker/FLPA; p205b ©Paul Hobson/FLPA; p206 ©Wolfgang Herath/Imagebroker/FLPA; p207 ©Derek Middleton/FLPA; p208 ©Paul Miguel/FLPA; p210 ©Olivier Digoit/Imagebroker/FLPA; p214-215 ©Robert Canis/FLPA; p216 ©Richard Becker/FLPA; p217 ©Marcus Webb/FLPA; p219 ©Paul Hobson/FLPA; p220 ©Bob Gibbons/FLPA; p221 ©Paul Hobson/FLPA; p223 ©Robert Canis/FLPA; p224bl ©IMAGEBROKER , BAO,Imagebroker/Imagebroker/FLPA; p224br ©;Reinhard Hölzl/Imagebroker/FLPA p225t ©Bob Gibbons/FLPA; p225b ©Hugh Lansdown/FLPA; p226t ©Jurgen & Christine Sohns/FLPA; p226b ©Robert Canis/FLPA; p227t ©Michael Krabs/Imagebroker/FLPA; p227b ©Gianpiero Ferrari/FLPA; p229 ©Erica Olsen/FLPA; p230t ©IMAGEBROKER , DBN,Imagebroker/Imagebroker/FLPA; p230b ©Ray Bird/FLPA; p231t ©ImageBroker/Imagebroker/FLPA; p231b ©Tony Wharton/FLPA; p233 ©Michael Krabs/Imagebroker/FLPA; p234 ©Paul Miguel/FLPA; p235 ©Richard Becker/FLPA; p236 ©Andrew Linscott/FLPA; p237 ©Erica Olsen/FLPA; p239 ©Paul Hobson/FLPA; p240 ©Nicholas and Sherry Lu Aldridge/FLPA; p241t ©Hugh Lansdown/FLPA; p241b ©Ottfried Schreiter/Imagebroker/FLPA; p242 ©Roger Tidman/FLPA; p243 ©Robert Canis/FLPA p245; ©Wayne Hutchinson/FLPA; p246 ©Phil McLean/FLPA; p247 ©Matt Cole/FLPA; p249 ©Gary K Smith/FLPA